How We Got Here

How We Got Here

A Slightly Irreverent History of Technology and Markets

Andy Kessler

Collins

An Imprint of HarperCollins*Publishers*

For my Dad,
who sparked my interest in technology

HarperCollins books may be purchased for educational, business, or sales promotional use. For information, please write to: Special Markets Department, HarperCollins Publishers, 10 East 53rd Street, New York, New York 10022.

First Collins paperback edition published 2005

Designed by Joy O'Meara

Library of Congress Cataloging-In-Publication Data
Kessler, Andy.
   How we got here: a slightly irreverent history of technology and markets/Andy Kessler.
      p.   cm.
   Includes index.
   ISBN 0-06-084097-8
   1. Technological innovations.   2. Capital market—Technological innovations.   I. Title.
HC79.T4K47   2005
338'.064—dc22   2005040407

05 06 07 08 09 DIX/RRD 10 9 8 7 6 5 4 3 2 1

Contents

Foreword

Talk about twisty-turny paths. I started life as an electrical engineer and ended up running a billion-dollar hedge fund on Wall Street. Like a pinball bouncing off of bumpers, I've been a chip designer, programmer, Wall Street analyst, investment banker, magazine columnist, venture capitalist, op-ed writer, hedge fund manager and even a book author. My mother thinks I can't hold a job. Friends suggest I know very little about everything and a lot about nothing. That's hard to argue with. But throughout it all, I learned over time to live by five simple creeds:

1. Lower prices drive wealth
2. Intelligence moves out to the edge of the network
3. Horizontal beats vertical
4. Capital sloshes around seeking its highest return
5. The military drives commerce and vice versa

I'm not entirely sure how I came up with this list, but it has worked. I've invested by it and have read and understood the news by it. It has helped explain the unexplainable and has helped me peer into the fog of the future.

I sat down with two different groups of people and tried to explain why these creeds are so valuable. The first group, engineering students who lived by math and science, were confused over how technology leads to business, even though advancing technology has driven and continues to drive most everything. The next group, business school students, was combative; they suggested that business and management skills trump economics. Maybe so.

So both groups were equally skeptical, and barraged me with questions like:

"How do you know they work?"

"How come we've never heard of these things?"

"Can you prove it?"

"What if you're wrong?"

"You're just making this up, right?"

To explain in 20 minutes what took 20 years to seep into my sinews, I'd have to walk them through the history of the computer industry, from the transition from telegraphs to gigabit fiber-optic networks and show how we moved from the Industrial Revolution to an intellectual property economy. And that's the easy stuff. Add money, and a diatribe turns into a dissertation. I'd have to explain how stock markets came into being, and insurance and the follies of gold standards. And then somewhere in this tale would have to be the link between military doctrine and commerce.

I needed to answer a few too many burning questions.

What is the history of the computer industry? Of the communications industry? Of the Internet?

Why does the U.S. dominate these businesses?

Didn't the British rule the last big cycle? What happened to them?

Why do we have money? What is it backed by? What was the gold standard all about? Do we still have a gold standard?

How did the stock market come into being and what is it for? Aren't banks good enough?

Who wins—money or ideas?

Does the military get its technology from industry or the other way around? Did anything besides Velcro and Tang come out of the Space Program?

Why does the U.S. have any industrial businesses left?

There are too many questions to answer. So instead, I wrote this primer. Enjoy.

I hate to admit it, but it was taxes that got it all started.

In 1642, 18-year-old Blaise Pascal, the son of a French tax collector, tired of waiting for his dad to come play a game of "le catch." Blaise's dad was what is known as a tax farmer, sort of a 17th-century version of a loan shark, threat of broken bones and all. Tax farmers advanced tax money to the government and then had a license to collect taxes, hopefully "harvesting" more than they advanced.

Elder Pascal was constantly busy calculating and tabulating his potential tax haul. To help him out, Blaise envisioned a mechanical device with wheels and cogs and gears and numeric dials that could sum up numbers to eight digits long. That's 10 million francs. Dad must have been a top tax guy.

Clockmakers were the high-tech folks of the era so Pascal built a model for his device by modifying gears and dials that he probably scrounged from clocks. The Pascaline fit in a brass box and was an amazing 17th-century device. The computer industry was on its way, albeit at the pace of a woozy escargot.

In 1649, King Louis XIV granted Pascal a patent for his odd device but it failed to effect much change over the next 45 years. Pascal, by the way, would contribute more than a mechanical calculator to this tale. He proved that vacuums exist; that one could measure pressure by inverting a tube of mercury; and as a vicious gambler, tried to figure how to beat the house and ended up inventing probability theory.

In 1694, a German, Gottfried Wilhelm von Leibniz, created a box similar to Pascal's but his could actually multiply. Leibniz used something called a stepped drum, a cylinder with a number of cogs carved into it, and gears that would engage a different number of cogs depending on their position. It was incredibly complex, which is why very few were ever built.

Inside a Pascaline or a Leibniz box were two simple elements needed to create the modern computer: Logic and Memory. The memory depended on the position of the mechanical dials. If the dial said 5, unless someone moved it, it would stay at 5. That's pretty simple memory.

Pascal's complex gears produced the arithmetic logic. If you add a 7 to the 5, the first dial was designed to show a 2 and then would kick the second dial ever so slightly to have it incrementally carry a 1 to the 10s column. Logic.

All this was rather slow, figure an addition every second if you were lucky. Plus, you had to write down the results every once in a while. But it beat ink and paper, in both speed and accuracy.

So Pascal got the computer business started, but with a whimper, not a bang. Logic and memory. That's it. My kingdom for Logic and Memory. So simple, yet so hard to implement.

Let's fast forward a bit. In 1880, 238 years after the Pascaline, the constitutionally mandated U.S. census took place as usual. The results were ready in 1887. Because of population growth, many feared the results of the 1890 census wouldn't be ready until well after the 1900 census! Big problem. But the solution was simple. A guy named Herman Hollerith came along and invented punched cards for the census, based on punched cards that a countryman of Pascal's named Joseph Jacquard had invented in order to program automatic looms. But Hollerith invented three devices—a puncher, a sorter and a tabulator that would read the census punched cards and keep a running count. The memory was the holes in the punched cards, and the logic was a set of mechanical gears and wheels that would keep a running count. It was complex, but it worked. He built 50 machines, each capable of tabulating 7,000 records per day, roughly a 10 times improvement over hand tabulation. The 1890 census quickly counted 62,979,766 U.S. residents.

Hollerith formed the Tabulating Machine Company in 1896. TBC changed its name to IBM in 1924, but wouldn't have electronic computers until after World War II.

While Hollerith was counting cards, Thomas Alva Edison attached a thin filament between two wires and got it to glow inside a vacuum bulb. Over time, carbon deposits from the filament would darken the glass bulb. In 1883, Edison worked on getting rid of the carbon and put a metal plate inside the bulb. He applied a positive charge on the plate, figuring it would attract the carbon. The carbon still sprayed around, but Edison noted that when he put a positive charge on the plate, a current would flow and if he put a negative charge on the plate, no current flowed. He named it the Edison effect (what else?) but promptly forgot about it. This tri-valve or triode would turn out to be the perfect device for a logic element and set the stage for the invention of real electronic computers.

I found a hot company with the most interesting story. It went public on July 4th at $25 per share. On its first day of trading, it jumped to $40, then $50. A month later, on August 10, it was trading at $280 and on August 11, it peaked at $310. The next day it fell to $212 and by the 15th it was down to $172, ending the year at $150.

Amazon.com? Internet Capital Group? Yahoo!? Guess again. The year was 1791. The stock was the Bank of the United States, set up by Alexander Hamilton in 1790 to help restructure the new government's $80 million of debt from the Revolutionary War and General Washington's bar tab. And you just won't believe it, but this hot IPO somehow ended up in the hands of 30 members of Congress, the Secretary of War and wealthy citizens. Some things never change. The Bank of the United States was signed into law in February of 1791. To set the tone for enduring government bureaucracy, it took five more months for the Bank to prepare for its initial public offering.

It wasn't so much a stock that was sold, as a subscription or scrip for ownership. According to the Museum of American Financial History, you paid $25 and had to put up another $375 by July of 1793, but you owned a piece of the Bank of the United States. This scrip traded on the streets of Philadelphia, which was then the nation's capital, right next to the cheese steaks stands. But these scrips soon began trading in New York, where the real money resided, with paper and pricing news traveling back and forth by stagecoach. What Hamilton saw was the need for a liquid market for government debt, so that later on, he could raise even more debt. He modeled the Bank of the United States after the Bank of England, which was able to borrow long-term debt and finance a navy to whip the French.

No one wanted to own a high-risk, illiquid IOU from a brand-new government of the United States. Investors were more willing to take the risk if they knew they could sell the scrip at some point. Of course some idiot top ticked it at

$310, just like some idiot in 2000 would top tick the NASDAQ at 5000. With risk and liquidity comes volatility.

Trading scrip on the muddy streets of a New York was no way to go through life. So on May 17, 1792, 24 brokers and merchants met under a buttonwood tree, which has since been replaced by a building at 68 Wall Street. Voila! They formed the first organized stock exchange in New York. They were hungry for action and someone had to move those bonds and scrip around. A stock exchange could not be much larger than someone's voice could carry, so they eventually moved indoors to a rented room on Wall Street. This group became the New York Stock & Exchange Board, and all sorts of bonds and other bank stocks began to change hands there. Alexander Hamilton got his liquidity and eventually so did every other venture that needed capital to grow.

In the middle of 1944, a squadron of B-29 "Superfortress" bombers took off from China. Their target was the Imperial Iron and Steel Works in Yawata, Japan, a major supplier of armaments for Japanese battleships and tanks. Imperial Iron churned out some two million metric tons of steel each year, a big chunk of Japan's wartime output. The coke ovens at the steel factory were a major target of the Allies.

A total of 376 500-pound bombs were dropped from these B-29s. Oddly, only one bomb hit anything—accidentally taking out a power station three-quarters of a mile away from the Imperial coke ovens.

The need for precision weapons would both directly and indirectly launch the digital revolution: Transistors in 1948, lasers and integrated circuits in 1958, packet switching in 1964 and microprocessors in 1970, and that was just the easy stuff.

Using Edison effect tubes and relays and other forms of logic and memory, scientists and engineers invented electronic computers to help win World War II. John von Neumann at the Moore School at the University of Pennsylvania designed the ENIAC digital computer, the birth mother of the U.S. computer industry, to speed up calculations for artillery firing tables for Navy guns. At the same time, Alan Turing and the British at Bletchley Park designed the Colossus computer to decipher Enigma codes. A host of electronic devices at Los Alamos helped speed up difficult calculations to control the reaction of uranium–235 for the atomic bomb.

It is the very pursuit of those weapons that created huge commercial markets, and vice versa. Lasers emerged as researchers cranked up the frequency of radar microwaves to avoid fog. Microprocessors were invented to create cheaper calculators. And now not only can missiles take out coke ovens, they can take out something as small as a Coke can.

Of course, there is more to this story than just card counters and light bulbs, buttonwoods and bombs. Somewhere in this mess are the lily pads of progress, the winding path to wealth and well-being, the DNA of our modern economy. It's a twisted journey.

By 2004, using much faster logic and petabytes of memory, a company named Google would perfect the business of searching for things and become one of the most profitable companies in the world. Google, started by two Stanford engineers, used 100,000 cheap computers, each doing billions of additions per second, connected via a global network using, among other connections, undersea fiber optics. Private investors, fattened from a hungry stock market, helped fund the company's meteoric growth. Its profits came from lowering the cost of search for its users, as well as from increasing the effectiveness of business to reach these users. Unlike in the time of poor Pascal, the computer and communications business already existed, with trillions in global sales. Google, as the expression goes, was built on the shoulders of giants. Knowing more about those giants and how they came into being can help us create more things to build going forward.

What changed in the interim? What did Google have that Pascal didn't? Jolt Cola and Nerf Guns are only a partial answer. What were all the incremental inventions over those 362 years that made Google happen? Who are the inventors and what were they thinking about?

It didn't happen overnight, nor is there an obvious trajectory to all this. The story is one of progress, mistakes, invention and innovation. Combine brains, money, entrepreneurs, stock markets, global conflicts, and competition and you end up with the most important thing: increased standards of living. All from Logic and Memory. Come and see.

So where to start? There are so many moving parts to this story and I've got to start somewhere. Just before World War II and the creation of the British Colossus computer and the American ENIAC might make sense, but then I'd miss all the components and reasoning that went into creating them. I think the best place to start is at the beginning of the last era, the start of the Industrial Revolution. There we pick up the stock markets and a money system. And the technology segues nicely. We get electricity, which helps create the tubes and relays that go into those first computers. So it's with the Industrial Revolution we start.

PART 1:

THE INDUSTRIAL REVOLUTION

Cannons to Steam

All it took was a little sunshine.

In 1720, the weather improved in Britain. No reason. The Farmer's Almanac predicted it. Crop yields went up, people were better fed and healthy. The plague, which had ravaged Western Europe, ended. Perversely, a surplus of agriculture meant prices dropped, and many farmers (of crops, not taxes) had to find something else to do.

Fortunately, there was a small but growing iron industry. Until the 1700s, metals like tin and copper and brass were used, but you couldn't make machines out of them, they were too malleable or brittle. Machines were made out of the only durable material, wood. Of course, wood was only relatively durable; wheels or gears made out of wood wore out quickly.

Iron would work. But natural iron didn't exist; it was stuck in between bits and pieces of rock in iron ore. A rudimentary process known as smelting had been used since the second half of the 15th century to get the iron out of the ore. No rocket science here, you heated it up until the iron melted, then you poured it out. Of course, heating up iron ore until the iron melts requires a pretty hot oven and to fuel it, a lot of charcoal, the same stuff you have trouble lighting at Sunday BBQ's. Charcoal is nothing more than half burnt wood but, as we all know, if you blow on lit charcoal it glows and gives off heat. So the other element needed to create iron is a bellows, basically, like your Uncle Ira, a giant windbag. Medieval uncles got tired really quickly cranking the bellows, so a simple machine, basically a waterwheel, was devised to crank the bellows, powered by running water. Hence, early ironworks were always next to rivers. This posed two problems, the iron ore came from mines far away, and after a day or two, the forest started disappearing around the mill and the wood needed for charcoal came from further and further away. It is unclear if ironworks sold their own stuff or if middlemen were involved. This raises the age-old question whether he who smelt it, . . . well never mind.

Meanwhile, the iron you would get out of the smelter was terrible, about the consistency of peanut brittle; the sulfur content was high, because sulfur is in most organic material, especially trees, and the sulfur from the charcoal blended with the iron ore. It seems that wood refused to play a part in its own obsolescence for machine parts.

This pig iron, and lots of it, was used for cannons and stoves and things, but it couldn't be used for screws or ploughs or a simple tool like a hammer, which would crumble after its first whack.

Iron makers evolved their process, and added a forging step. If you hammered the crap out of pig iron, reheated it, and hammered it again, you would strengthen it each time until you ended up with a strong substance aptly named wrought iron. Besides gates and fences, wrought iron worked reasonably well for swords and nails and screws. But to create decent wrought iron, you really had to get the brittle out of the pig iron.

In 1710, Abraham Darby invented a new smelting

process using coke, basically purified coal, instead of char-coal. The resulting pig iron was better, but not perfect. His son, Abraham Darby II improved on his dad's process and by mid-century, was oinking out pig iron usable for wrought iron, but only in small quantities. Unfortunately, like iron ore, coal was far away from the river-residing ironworks, so roads were built (sometimes with wood logs) and wagons brought coke to the river works. No surprise then that many ironworks moved to be near the coke fields. In 1779 Darby III would build the famous Ironbridge over the river Severn to transport materials with the iron supplied by the process his father and grandfather had developed. Heck of a family.

Demand for iron ore and coke took off and mining became a big business. One minor problem though, mines were often below the waterline and flooded constantly. This cut down on dust but too many miners drowned, hence the huge demand for something to pump out that water. The answer was a steam engine.

The concept of an engine run by steam had been around since the ancient Egyptians. It is not hard to image someone sitting around watching a pot of water boil and remarking that the steam coming off expands, and thinking, "Gee, if I could just capture that steam, maybe it would lift that big rock to the top of that pyramid." In fact, an Egyptian scientist named Hero living in Alexandria in 200 BC wrote a paper titled "*Spiritalia seu Pneumatica,*" which included a sketch of steam from a boiling cauldron used to open a temple door. It looked like a failed seventh grade science fair project.

Not quite a couple of thousand years later, steam proj-ects started boiling up again. The most obvious contraptions were high-pressure devices, with which you boiled water, generated steam in a confined location, and the increased pressure would move water through a pipe, which might turn a waterwheel, or turn gears, or even just operate a water fountain. The problem was that in the 17th century, materials for the boilers were a bit shoddy, and most experiments ended with boiler explosions, a nasty occupational hazard.

Most of what I learned about steam engines was from reading Robert Thurston's book titled *A History of the Growth of the Steam-Engine* which he published in 1878, but still remains an invaluable resource in understanding the subtleties of this new invention.

Back in 1665 Edward Somerset, the second Marquis of Worcester (but he tried harder) was perhaps the first to not only think and sketch a steam engine, but also build one that actually worked. He created steam in a boiler, and had it fill a vessel already half filled with water. He then had the steam run out to another cooler vessel, where it condensed back into water. The lower pressure of the escaping steam would create a vacuum that would suck water into the first vessel to replace the steam that left. Unfortunately, Somerset's engine was only good at moving water. It operated fountains, but had very few other applications.

In 1680, the philosopher Huygens, who gets credit for inventing the clock, conceptualized the gas engine. Gunpowder, he figured, exploding inside a cylinder could push a piston up. The explosive force would expand the gas and lift the piston, and would remove all of the air from the cylinder through a set of open valves. The valves would then be closed, and the subsequent vacuum would pull back down the piston. It didn't work, but it introduced to the world the idea of a cylinder and a piston. You have a bunch of them in your car, with gasoline replacing the gunpowder. And instead of a vacuum pulling down the piston, you have an explosion in an adjacent cylinder mechanically move the piston back down. That's why you have a 4- or 6- or 8-cylinder engine in your car, or for real dynamite starts from red lights, 12 cylinders.

One of Huygens's students was the Frenchman Denis Papin, a Protestant who left France when Louis XIV decided he didn't like Protestants. He ended up in London where in 1687, he invented the "Digester" pressure cooker. This was a sealed pot with a safety valve on top that opened when pressure got too high, cutting back on explosions. That valve was a critical addition to the evolution of a useable steam engine. Papin then wandered over to Italy and Germany and in 1690, came up with a steam engine by modifying the Huygens design. He filled the bottom of the cylinder with water. A flame heated the water to a boil, which created steam. The steam would lift the piston up. Then he removed the flame. The steam condensed (notice Papin didn't do anything but remove the flame), forming a vacuum, which sucked down the piston, and it started all over again. It worked. His cylinder was 2½ inches in diameter and could lift 60 pounds once a minute. Big deal, you and I could lift that much all day. But he figured that if the cylinder was 2 feet in diameter and the piston 4 feet long, it could lift 8,000 pounds, 4 feet, once a minute, which was the power of one horse.

Now we're getting somewhere.

Papin never built the bigger model, and when he started telling people about his new invention, the steamboat, local boatmen heard about it and broke into his shop and destroyed it. They (correctly, but early) figured it would threaten their full employment. This destruction will be a recurring theme.

Thomas Savery of Modbury was a mathematician and a

mechanic who was familiar with the works of both Somerset and Papin. He took the Marquis's two-vessel design, and
added a useful cock valve to control the flow of steam
between the two, and then three, vessels. He also ran some
of the pumped water over the outside of the vessels to create
surface condensation, which helped the steam condense and
the engine run faster. And thus he produced what he called
the Fire Engine.

In July of 1698, he took an actual working model of the
Fire Engine to Hampton Court to show it to officials of King
William III. He was awarded a patent:

> A grant to Thomas Savery of the sole exercise of a new inven-
> tion by him invented, for raising water, and occasioning
> motion to all sort of mill works, by the important force of fire,
> which will be of great use for draining mines, serving towns
> with water, and for the working of all sorts of mills, when they
> have not the benefit of water nor constant winds; to hold for
> 14 years; with usual clauses.

Those usual clauses were probably kickbacks to the
King's Court, but Savery had 14 years to run with his new
engine. I'll get to the beginning of laws for patents in a bit.

He marketed it as the Miner's Friend. Miners were using
horses, as few as a dozen to, in some cases, 500, to pull up
full buckets of water—the old bucket brigade. A device that
burned wood or coal and pumped water was a gift from
heaven.

A few miners used the Savery Fire Engine, but for depths
beyond 40 or 50 feet, the suction was not enough to pull up
much water. After Savery died in 1716, a man named Jean
Théophile Desaguliers took up where he left off. To generate
more vacuum, he collapsed the design down to one vessel,

or receiver, and invented a two-way cock that would allow steam into the receiver when it was turned one way, and would allow in cold water to condense the steam when it was turned the other way. He also turned the incoming water stream into little droplets, which accelerated the condensation and created the vacuum faster. But when the cock was turned toward letting in cold water, the boiler would fill up with stream, at high pressure. Developing it, Desaguliers probably killed quite a few apprentices and workers with exploding boilers.

Measurements in 1726 showed this design capable of the power of 3 horses. And you didn't have to clean up after them.

Still, as a useable tool, even for pumping out mines, it was lame. But demand was there. The bucket brigade was replaced with a pump, basically a vacuum generated by a rod moving up and down deep in the mine, powered by a windmill or lots of horses.

Fifteen miles down the road from where Savery hailed, in Dartmouth, a blacksmith and ironworker by the name of Thomas Newcomen thought he could come up with a steam engine for the nearby mines. It appears he had seen the Savery engine, and must have either seen or heard about Papin's design. What Newcomen did was combine the best of both, the Savery surface condensation vessel design with the Huygens/Papin cylinder and piston design, to create, in 1705, an "Atmospheric Steam Engine."

A boiler would feed steam into a cylinder, until the piston reached the top. Then a valve was turned to cool the outside of the cylinder or, in improved designs, add droplets of cold water inside the cylinder. The steam would condense, create a vacuum, and pull down the piston. Instead of pumping water directly with that vacuum, it would move a beam above it up and down. A pump rod attached to the beam would operate a water pump in the mine.

Newcomen had a small legal problem, in that the Savery patent seemed to cover any steam engine that used this surface condensation method. So in 1708, the two men struck a deal to co-own the patent. In this way, Savery managed to cut himself in on the lucrative steam engine market even though his own design never really worked.

The combination worked wonders. A two-foot diameter piston operated at six to ten strokes a minute. Then, a young boy/wizard named Humphrey Potter added a catch so that the beam moving up and down would open and close the valve to let in the condensing water, and the speed cranked up to 15–16 strokes per minute. Conceptually anyway, it could pump 3,500 pounds of water up 162 feet. That's the power of eight horses. At a stroke every four seconds the Newcomen steam engine must have been an amazing sight in its day.

The Newcomen engine was the prevailing design for the first half of the 18th century. Miners bought it to pump out their floods. Some low-lying wetlands were pumped out. Some towns even used it for their water supply. But in reality, it was not a huge success. The Industrial Revolution didn't start until late in the 18th century. You might say that Savery and Papin each released version 1.0. Newcomen combined them and released version 2.0. But it wasn't enough. Where was 3.0?

In 1774, the Iron Master of Shropshire, John Wilkinson had a serious problem. He had a backlog of orders for cannons from King George, who was trying to put down those pesky colonists in the New World. Wilkinson desperately needed a source of power to operate his bellows to smelt iron ore to pour into cannon casts.

He stumbled on the solution while watching a funky new steam engine pumping out his own flooded coal mines. This almost 3.0 steam engine would have a profound influence on industry, but that wasn't so obvious at first.

It was, of course, James Watt's steam engine, but it still wasn't all that good. Back in 1763, James Watt was employed at Glasgow University, with the task of fixing a Newcomen steam engine. Fifty years after Newcomen's invention, five horsepower was still not very efficient, plus it broke down all the time. And, someone had to constantly seal the cylinder to prevent the steam from leaking out and the vacuum from weakening. To give you an idea how rudimentary this was, the sealant usually took the form of wet ropes.

Like all good engineers, Watt took it apart to figure out how it worked. He noticed that the biggest problem with the Newcomen engine was that because it kept blasting cold water on the outside and inside of the cylinder, it wasted as much as three-quarters of the energy used to create the vacuum. And it took time for the cylinder to heat up enough to accept new steam without instantly condensing it.

His professor at Glasgow University, Dr. Black, had been teaching courses for two years on theories regarding latent heat. Adam Smith was a professor at U of G around the same time, and in fact Smith and Black were good friends. Latent heat is the reason you put ice cubes in your soda. No matter how much heat is applied by the hot sun at a baseball game, for example, all the ice has to melt before the soda increases in temperature. Latent heat means you can add heat to a pot of water, but it won't boil and give off steam until the entire pot of water is at 212 degrees Fahrenheit. In other words, he sort of proved that a watched pot never boils.

Watt ran a series of experiments to measure temperature and pressure and proved a prevailing theory, that steam contained "latent heat." It's nice to have a smart professor as your mentor, and perhaps this is an early example of a technology spinout from universities. Watt theorized that the cylinder had to be as hot as possible, boiling hot, before new steam added to it would stay steam and not condense. Off went a light bulb in his head.

He would later write:

> I had gone to take a walk on a fine Sabbath afternoon. I had entered the Gree by the gate at the foot of Charlotte Street, and had passed the old washing house. I was thinking about the engine at the time, and had gone as far as the herd's house, when the idea came into my mind, as steam was an elastic body, it would rush into a vacuum, and, if a communication were made between the cylinder and an exhausted vessel, it would rush into it, and might be there condensed without cooling the cylinder.

Happens to you all the time, right? But he was right about steam being elastic, in more ways than he even knew!

So Watt added a modest improvement to the Newcomen design. He created a separate chamber outside of the cylinder. This "condenser" was kept underwater, as cool as possible. At the top of the piston's stroke, a valve would open and the hot steam would be allowed to flow out, into the chamber, where it would condense into water, creating a vacuum, and thus pulling down the piston. The cylinder and piston would stay hot, so when steam was added back in, it would quickly fill the cylinder and be ready to flow out again into the condenser.

At least that was the concept. Most stories of Watt stop here, but it gets better.

In reality, between 1763 and 1767, Watt literally went broke trying to perfect his new design. It leaked steam like crazy, as the cylinders were not "true." So his new condenser was not very efficient. Watt even had to get a real job as a surveyor to support his inventing. In 1767, John Roebuck, the owner of a Scottish iron foundry, assumed Watt's debts up to 1,000 pounds, and gave him fresh money to improve his design. In exchange, Roebuck received two thirds of any patent.

In 1769, Watt was granted a patent for his steam engine design by Parliament, which had taken over this duty from the King. Of course, Parliament was run by property owners who, not surprisingly, were all for upholding property rights.

Almost simultaneously in 1769, old John Roebuck went bust, and fell under the burden of his own debt. Fortunately, there were other businessmen around who understood the need for engines to pump water and maybe even to drive factories. These were called the Lunatic Fringe.

Dissenters were out on the Fringe. As outsiders, no one would hire them, so they were forced to be entrepreneurs. A famous group of Dissenters, mainly Quakers, started meeting in 1765 in homes around Birmingham, to discuss the latest scientific trends and the latest in thinking. They would meet each month, on the Monday night closest to the full moon. There were no street lights back then, and clearly no *Monday Night Football*.

They called themselves the Lunar Society, although it was nicknamed the Lunatics by the butler of one of its members, Samuel Galton Jr., a banker and gun maker. No question this group was the lunatic fringe, but by pursuing the latest in scientific advances, they were perhaps saner than the rest of England. I could describe a few here, but you'll see Lunatics pop up again and again in this story.

Matthew Boulton was born into the silver stamping business, but he had struck out on his own in 1762 as a manufacturer of luxury goods, also known as a "piecer." He created the Soho Manufactory a couple of miles outside of Birmingham and made buttons, buckles (Puritans loved buckles), vases, statues and low-end clocks. Constantly on the lookout for ideas and processes that could improve his shop, you might say he was the venture capitalist of his time.

Boulton was buddies with Ben Franklin and the two often corresponded about steam and steam engines, even about one of their own. Franklin was probably trying to find a new market for his potbelly stove!

In 1768 meanwhile, Watt stopped by the Soho Manufactory and finally met Boulton. They discussed his new engine as well as its potential uses in the factory and even for driving carriages. It was a fateful meeting, because a year later, Watt needed to raise money fast to buy out his defunct partner.

Boulton agreed to buy out Roebuck's two-thirds interest in the patent, but more important, to fund Watt's continued research to improve his external condenser steam engine and make it work.

The plan was to move Watt down to Birmingham, because that was where the demand was, but Watt had some unfinished business in Scotland. He kept toiling over his condenser, and didn't journey south until 1774.

The Newcomen design was still selling, despite all its flaws, but demand was strong for more powerful engines. Watt's biggest problem was getting the materials and labor he needed to accurately construct his engine. The cylinder was key, and it's "trueness," how accurately round the cylinder was, directly affected its power by stopping leaks. Watt was excited when he constructed a cylinder that was within ⅜ of an inch of being a "true cylinder." Doesn't sound like much, but to steam, ⅜ of an inch is as wide as a country mile.

The London Water Works approached Boulton about pumping engines to help supply the city with water. He jumped into action, but only to protect his investment. In 1775, he went to London, and had a few influential friends introduce legislation in Parliament to extend Watt's patent, which was set to expire in 1783. Fellow Lunar Society member Joseph Banks was president of the Royal Society and had great connections. The bill passed and Boulton and Watt had the patent on Atmospheric Steam Engines (with cool condensers) until 1800, even though it still didn't work that well.

The Brits were at war on and off with the French for centuries, but never more so than in the 18th century over territories in the New World, especially in Canada over fur, in the West Indies over sugar, cotton and rum, and over trade routes to central Asia. Add to that those rebellious, thankless colonists from Massachusetts to Georgia.

King George needed cannons for his troops in the American Colonies as well as for his warships to keep the French on their side of the English Channel. Back to John Wilkinson, the Iron Master of Shropshire, who had a precision cannon-boring tool, which was in essence a big monster lathe. This tool cut true, making highly effective cannons with ever so narrow "windage," the gap between a cannon's barrel and the cannonball. The smaller the windage, the greater the distance the gunpowder's blast would propel the cannonball, rather than leak out past it. Wilkinson's cannons shot cannonballs farther and higher and faster.

Wilkinson chose not to patent the tool, since he didn't want to generate drawings that others could copy. Based on this tool and his reputation, Wilkinson had a huge backlog of orders from the King for cannons.

But Wilkinson (not to be confused with the sword dude, James Wilkinson as is often the case) had recently moved his foundries from river's edge to the coalfields. He switched from wood to coal, but lost his source of power: the river.

You see, sulfur-rich charcoal from half-burnt wood and power from waterwheels just didn't cut it for making cannons; the pig iron it produced was too brittle. High-grade coal or coke was plentiful up in the hills of Staffordshire and contained almost no sulfur, so the iron was sturdier. Wilkinson solved one problem but ended up with another: how to operate his bellows, to blow enough air to heat up the coke to an intense enough heat, so he could smelt iron ore, and pour the liquid iron into casts of cannons. His boring tool also needed a source of power, waterwheels were slow and horses expensive to feed.

Wilkinson needed piping hot coals to smelt his iron. He had seen James Watt's steam engine pumping water out of coal mines, and thought he could use it to run his bellows rather than horses. So Wilkinson went through the hassle of moving it and attaching it to his bellows, but was terribly disappointed. The cylinder was awful. It leaked steam with every stroke, robbing the engine of most of its power.

So just as Watt took apart Newcomen's engine to figure out how it worked, Wilkinson took Watt's engine apart, and probably started laughing. Peeling away the wet hemp used as a sealant, he noticed Watt's cylinder looked as craggy as the coastline of England.

While Watt was proud of his ⅜ of an inch from true cylinders, Wilkinson had his secret boring tool and could make Watt's cylinders 10 times more true. So he ripped the engine apart, recast the leaky cylinder using his precision boring tool and found it generated 4 to 5 times more power, enough to run his bellows. This meant 25 to 40 horsepower engines, up from 5 to 8. And that was the push the world needed to take it into the industrial age. Sometimes it's that simple, and easy to miss.

Being a reasonable businessman, he told Boulton and Watt that he could improve their crappy little steam engine by a factor of five, in exchange for the exclusive rights to supply precision cylinders to B&W.

Deal.

Here is the part of the story that may sound familiar to today's technology business. With flooded mines (market demand), Watt's condenser (technology), Boulton's money (capital), Parliament's patent (intellectual property rights), a ready workforce (religious persecution), and Wilkinson's precise cylinders (technology), they had just about everything. What they were missing was a successful business model. It was Matthew Boulton who came up with one.

Boulton and Watt didn't actually sell steam engines since no one could afford one. Most of the early customers were Cornish mines. Beyond a few Parliament-sponsored joint-stock companies, the stock market and banking system were not quite developed, especially for risky businesses. Limited liability for corporations wouldn't be the law until 1860.

Miners lived day to day and used a cost book system of accounting (I slept through accounting, too). At the end of each quarter, all the partners in the mine would meet at the Count House to go over the numbers and split any profits. These Count Dinners were giant drunk fests, each mine vying for the prize of offering the most potent Hairy Buffalo punch. At the end of the night, the mine companies were drained of cash and the miners drained of brain cells.

So instead of selling steam engines, Boulton just traveled around to mines (and later mills and factories) and simply asked the miners how many horses they owned. Boulton and Watt would then install a steam engine, and charge one third of the annual cost of each horse it replaced, over the life of the patent, that is until 1800. Back then, a horse cost about 15 pounds per year, and I have seen figures that showed that components for a 4 horsepower engine would cost 200–300 pounds. A 50 horsepower engine would cost around 1,200 pounds.

Not having four eating and defecating horses around could save someone 60 pounds a year and Boulton and Watt would charge just 20 pounds a year. Paybacks for the miners were immediate. And if the 4 horsepower engine cost, say 200 pounds, B&W started turning a profit after ten years. Or they could charge 250 pounds a year for a 50 horsepower engine and turn a profit in less than six years. And they had a 25-year patent.

It was in Boulton and Watt's best interest to install more powerful engines—they just needed to find something beyond pumping out flooded mines to drive demand for horsepower. A barrier lowering the cost of power had just been busted down, but it wasn't a gusher.

So did the Industrial Revolution start then and there? Not quite yet. With much of his profits Watt continued tinkering with his engine. In 1781, he patented the "sun and planet wheels." The planet gear was connected to the overhead beam of the steam engine and went up and down. But its teeth were attached to the sun gear, which was attached to a big wheel. The up and down motion of the steam engine was transferred to a circular motion of the wheel, something that was critical, if you wanted to grind flour or operate machinery.

A year later, Watt paid Wilkinson back for his precision cylinder. He did this by inventing a steam hammer, or "tilting forge hammer," for the forging step, to beat the pig iron and turn it into strong wrought iron. The 120-pound hammerhead rose eight inches and struck 240 blows per minute. Bet you couldn't build one of those in your basement today.

Also that year, Watt developed the double acting steam engine, which alternately fed steam into each end of the cylinder, kind of a push-me-pull-you contraption. But now his steam engine applied power in both directions of movement, thereby increasing its power.

The steam engine made smelting faster and better, but despite a steam hammer, the forging stage was still much too slow. Around 1784, Henry Cort perfected the forging process, by getting rid of the nasty hammering. Instead, he had the liquid iron stirred in a big pot and then poured between rollers as it cooled. Puddling and rolling was such an obvious improvement, in hindsight, that productivity shot through the roof since it cut forging times for a ton of pig iron from 12 hours to 45 minutes.

It wasn't until John Wilkinson's combining of the Darby coke smelting process, the Boulton and Watt steam engine, his own precision cylinders, and the Cort forging process, that we got industrial strength iron, and lots of it. Boulton and Watt brought down the cost of power, probably by a factor of 10.

The religious dissenters, the Lunatic Fringe, were the day's factory owners. A Corporations and Test Act barring them from public office was in force. But as wealthy businessmen, they fought for more political power. For example,

Wedgwood, the pottery maker, and Boulton founded the General Chamber of Manufacturers in 1785 in order to gain more clout. Along with Joseph Priestley, a member of Birmingham's Lunar Society and a friend of Franklin's, they lobbied for repeal of the Corporations and Test Act, although that wouldn't happen for another 43 years. They had to prove themselves first.

Did we get anything we need for the computer business? Lots, but most of it came indirectly. That steam engine would be indispensable for generating electricity, but not for a few more decades. Money, and keeping track of money, would become much more important as well. But, like the early computer business, the steam engine needed some other market beyond pumping out mines or puffing bellows for this whole thing to work.

By 1800, 500 steam engines were up and pumping, and the economic engine was lifting off. Something important had changed.

While iron is nice, clothes were a much bigger market in the late 18th century. But individuals did carding, spinning and weaving at home. Automation was never thought possible, but as soon as new human-operated tools and machinery came about, the need for power intensified.

Carding, usually done by children, involved pulling fine clumps of fiber, 12 inches long and an inch thick, from a chunk of wool or cotton using hand cards—wooden blocks with metal spikes. The cardings were then fed into a foot-powered spinning wheel, called a spinner, where it was turned into thread or yarn. This was then weaved on a loom to create cloth.

Looms have been around forever. American Indians, the Greeks, the Egyptians, all used looms to weave clothing. A loom is nothing more than a wooden frame. You would warp a loom by winding thread or yarn up and around the top and bottom frames, making a figure eight each time. You would weave by taking thread or yarn, called weft, and moving it with a shuttle sideways through a gap in the warp called the shed. Then you would move a sword to tighten the weft, and move the sword toward the other end of the warp to send the shuttle through the new shed. Got it? Makes steam engines sound simple.

But looms are actually very simple but extremely labor intensive. A weaver must pay careful attention. In 1733, John Kay patented a wonderful device called the flying shuttle. The loom had a long box known as a shuttle race attached to it. With a set of cords rigged above the loom, a weaver could send the shuttle back and forth with ease, and with one hand, so with his other hand he could control the thread or drink some tea. Foot pedals moved the warp up and down to create the shed.

Productivity popped. Weavers became faster and could weave longer materials. But it doomed the home-based textile industry. Not surprisingly, in 1755, a mob broke into John Kay's house and destroyed one of his looms, but fortunately, he had a few spare.

Making thread or yarn, on the other hand, remained old-fashioned. Sheer a sheep, and then wind the wool on a spinning wheel. Invented who knows how many centuries before, as quaint as could be. But as weavers demanded more yarn of higher quality, they substituted cotton from the New World for expensive wool.

Along came the Spinning Jenny. Invented in 1764 by James Hargreaves, it combined eight (and eventually 80) spindles of wool into a thread strong enough to sew with. Hargreaves got the idea (or at least according to legend) when his daughter Jenny knocked over the family spinning wheel and had to chase it through the house. When local spinners heard of the invention, they broke into his home in Lancashire and busted the Jenny up, the wooden one. He moved to Nottingham.

Around the same time, Richard Arkwright, a wigmaker, was also spinning. He hoped to create yarn, which required winding together threads into a thick bundle. He hired a clockmaker, as they were the engineers of their time, to help him put together a machine. They created a Spinning Frame, which combined various spindles that twisted the threads at different speeds. Originally, Arkwright thought that spinners would operate the frame by cranking a handle. No dice. This thing was a bear. He ended up needing horses to operate it, and even that proved insufficient, so he moved the whole thing riverside, where a waterwheel and fast-running water could drive the machine, without getting either it or the yarn wet. It became known as a Water Frame. If ever there was a need for an engine, it was the Water Frame, and indeed, around 1785, Arkwright hooked up Watt's engine and created a new engine market, one much bigger than for mines and ironworks.

Jedediah Strutt was a wealthy businessman in Derbyshire who ran a hosiery business. In London, he would buy raw silk from India and employed local workers to weave it into high-end socks. Of course, silk was quite expensive, and he figured he should move into wool or even better, cotton socks, because they were cheaper to make and were, perhaps, much larger markets. He met Arkwright and in 1771 put up the money to build a Water Frame in Cromford on the River Derwent.

Arkwright did it all. In 1775, he also patented a Carding Engine, a spinning cylinder with a comb that moved up and down to help automate the task of generating cardings that are fed into the spinning frame. With his Water Frame churning out yarn, he built cottages near the factory for literally thousands of workers who weaved them into cloth and clothing, thereby establishing the original cottage industry. Two-thirds of the workers in the Water Frame factory were children. Arkwright would never hire five-year-olds, he felt it was prudent to wait until they were six! No one over 40 could get a job there. It wasn't until the acts of 1802,

1819 and 1831 that anyone worried about working conditions.

So important was Arkwright's factory to the well-being of Derbyshire, that 5,000 men there developed a network of signals to respond to any mobs then threatening to destroy the Water Frame, which was endangering the home-based spinning industry. In 1779, in a village near Leicestershire, Ned Lud led a group of workmen and broke into the home of a stockinger, a socks maker. They destroyed several stocking frames, considering the frames a threat to their jobs as artisans. These so-called Luddites were not the first and certainly not the last group to feel threatened by automation. But all you have to do is look at Arkwright's Water Frame factory and cottages to figure out that automation creates plenty of new jobs.

Still, the yarn from a Water Frame was thick and the thread from the Spinning Jenny was coarse. One can only imagine how itchy clothing was in 1775, not just clothes from wool but cotton as well. Royalty still insisted on silk, it beat scratching and twitching all day. Comfortable clothing was yet another thing that separated the rich from the poor.

Tailors were interested in a yarn that was strong, smooth and soft, to replace expensive silk. In 1775, an inventor named Samuel Crompton crossed the Jenny and the Water Frame and invented the Spinning Mule. This machine didn't just spin or twist, its spindles moved back and forth up to 5 feet. This effectively stretched the yarn to "silky smoothness," and then quickly wrapped it into a bobbin as it loosened.

By 1790, 400 spindles hung off the Spinning Mule, and no man or mule or horse or even running water could keep up with the power needed to run one of these things. Boulton and Watt steam engines to the rescue. The Spinning

Mule was a breakout device. It was just what the textile business needed: cheap, smooth material. And of course, it was just what Boulton and Watt needed, something to soak up lots and lots of horsepower.

The Carding Engine stripped the fibers into cardings. Spinning Jennies created thread. Water Frames created yarn. Spinning Mules turned out smooth yarn and thread. Looms were still run by hand. So, around the time Watt was extending his steam engine patent 25 years to 1800, all but the weaving step of textile manufacturing was under mechanical power that steam engines could create.

In 1785, Edmund Cartwright sought to fix this problem by applying mechanical power to hand looms. But first he had to wait for Arkwright's patent on cotton spinning to expire. He knew that cotton mills would then be built, which would turn out an abundance of thread and yarn. Cartwright thought about starting his own cotton mill but figured, smartly, that everyone and his brother would start one of those. Instead, he wanted to leverage the abundance of yarn, not help create it. So he began working on a Power Loom. He was only two hundred years ahead of his time in innovative business thinking. The world would eventually catch up.

Without even looking at a hand-operated loom, he built a fully mechanical one and patented it in 1785. It was worthless. But he persisted and eventually he fully emulated the hand and foot movements of weavers with mechanics. According to the Supplement to the *Encyclopedia Britannica*, here is Cartwright's own account:

Happening to be at Matlock, in the summer of 1784, I fell in company with some gentlemen of Manchester, when the con-

versation turned on Arkwright's spinning machinery. One of the
company observed, that as soon as Arkwright's patent expired,
so many mills would be erected, and so much cotton spun,
that hands never could be found to weave it. To this observation I replied that Arkwright must then set his wits to work to
invent a weaving mill. This brought on a conversation on the
subject, in which the Manchester gentlemen unanimously
agreed that the thing was impracticable; and in defence of
their opinion, they adduced arguments which I certainly was
incompetent to answer or even to comprehend, being totally
ignorant of the subject, having never at that time seen a person weave. I controverted, however, the impracticability of the
thing, by remarking that there had lately been exhibited in
London, an automaton figure, which played at chess. Now you
will not assert, gentlemen, said I, that it is more difficult to
construct a machine that shall weave, than one which shall
make all the variety of moves which are required in that complicated game.

Some little time afterwards, a particular circumstance
recalling this conversation to my mind, it struck me, that, as in
plain weaving, according to the conception I then had of the
business, there could only be three movements, which were to
follow each other in succession, there would be little difficulty
in producing and repeating them. Full of these ideas, I immediately employed a carpenter and smith to carry them into
effect. As soon as the machine was finished, I got a weaver to
put in the warp, which was of such materials as sail cloth is
usually made of. To my great delight, a piece of cloth, such as
it was, was the produce.

I wonder if the chess machine ever won.

Cartwright opened a weaving mill in 1787 in Doncaster,
where workers simply fed in or fixed broken thread. He

tried to use a waterwheel to operate the weaving mill, but quickly contacted Boulton and Watt and hooked up a steam engine.

An experienced hand weaver could, according to Richard Guest's *Compendious History of the Cotton Manufacture 1823,* produce "two pieces of nine-eighths shirting per week; each twenty-four yards long, and containing 105 shoots of weft in an inch." A young kid overseeing a Power Loom, run by coal and steam, could produce one each day. That's 3.5 times productivity.

As you can imagine, the mechanization of cloth making from start to finish brought the cost down considerably. This also meant a huge increase in demand for raw cotton.

Despite an annoying interruption of the American War of Independence, the Revolutionary War, or as it was known in England, the War with Those Damn Colonists, the Triangle Trade picked up steam. Finished materials went to the coast of Africa, traded for unfortunate slaves who were transported to sugar and tobacco and cotton plantations, and traded for the raw materials needed as inputs to the new Industrial economy.

Cotton was hot. Operators of Spinning Frames and Power Looms were demanding more and more raw cotton from the New World.

The long staple variety of cotton was brought from the West Indies to Georgia and South Carolina. The good news was that its seeds were really easy to remove from the "cotton boll." The bad news was that it needed a long growing season and humidity and so only grew in coastal regions. In the early 18th century, that was not a problem. Land was so plentiful that farmers didn't bother with fertilization, and soil exhaustion and erosion was commonplace. If the fields no longer produced cotton, they just planted another field.

But ever-rising demand from Spinning Frames and Mules and Weaving Mills sent farmers inland chasing vaster fields.

Only the short staple variety of cotton grew inland, and it had green sticky seeds. Slaves were used to remove the seeds from the cotton bolls, which were then shipped to England. It took a person, usually a slave, a full day to remove those sticky green seeds from one pound of cotton.

A Yalie named Eli Whitney was hoping to start a law career, but had to pay off his student loans. So in 1792, he headed south, and ended up as a private tutor on a plantation in Georgia, owned by Catherine Greene. The South was in turmoil as tobacco profits were disappearing from oversupply, and cotton was the only salvation. But it was impossible to make a profit with such a tedious problem of separating seeds.

It didn't take Whitney long to figure out that Greene, and every other plantation owner, needed a machine to remove those damn sticky seeds. He convinced Greene to set up a laboratory and in the winter of 1792 he created the cotton gin. In case you were wondering (because I was), gin was short for engine, in good old Georgia talk.

The gin was a device with spiked teeth on a rotating cylinder that pulled the cotton through small slots, leaving the seeds behind. Then a brush came along, rotated by pulleys and belts, and pulled the cotton lint off the spikes. Operating it with a hand crank, someone could clean 50 pounds of cotton a day instead of just one, perhaps helping date the phrase, "Now just one cotton-picking minute."

In 1794, Whitney patented his cotton gin.

Every plantation owner needed one. Whitney took on a partner, Phineas Miller, and like Boulton and Watt, came up with a unique business model to leverage the invention.

Instead of selling cotton gins, they made a whole batch of them, and ginned the cotton for the plantation owners for a fee of 40% of the profits, paid in cotton. That was pretty close to the one-third savings on horses that B&W charged.

Of course, U.S. patent law was pretty new, and most farmers just made their own gin, and if anyone asked, said that theirs was a new invention, better than Whitney's gin. That's where Miller proved his worth as a partner. He sued everyone he could find who copied the gin. But the United States Patent Act of 1793 was vague, and it wasn't until its wording was changed in 1800 that Whitney and Miller were able to win any suits.

Meanwhile, Whitney built bigger and bigger cotton gins. Waterpower replaced the hand crank or horse. America didn't have any steam engines, and it would be a while until that mode of industrialization took hold. But it did have the ability to turn out more cotton. In 1792, no more than 150,000 pounds of American cotton made its way to England. Eight years later it was 17 million pounds. In 1850, 700 million pounds of cotton were exported to England, all to be run through the Frames and Mules and Looms. The Industrial Revolution was on!

Many thought the invention of the cotton gin would help end slavery, as that labor-intensive task was mechanized and productivity increased by a factor of 50. But as you can see from the cotton export numbers, business boomed, actually increasing the demand for slaves. The Triangle Trade grew larger, as more slaves were needed to pick more cotton every year. It's hard to believe, but the spindle cotton picker that automated that task wasn't invented for another century and a half, finally showing up in 1940.

With steam-powered mills and a steady flow of clean cotton, England now had the economic engine it needed. Import raw materials like cotton, run them through industrial machinery run by steam-powered engines, and export finished goods like yarn and cloth and textiles, at prices so much cheaper than handmade goods, it changed the way the world dressed. And in less itchy clothes at that. Mechanized threshers helped bring the cost of food down. Iron became cheaper, as did anything made from iron: pots, pans, kettles, ploughs, and screws. Lunar Society member James Keir, with cheap ironworks available, set out to create new metals, and invented a bolt metal strong enough to bolt parts of ships to each other tightly and useful in making window frames. The British Navy was a big user of these bolts since its ships, armed with Wilkinson cannons, protected this booming trade.

Industrialization was not some master plan to remove workers from their century-old tasks; instead it was a complete reengineering of life based on the ability to lower the costs of daily staples. This price drop would repeat itself through history as a sign of maximum change, increased standards of living, and turmoil.

Steam, cannons, looms, windage? What does this have to do with computers and communications and stock markets and innovation?

OK, OK, just a few more piece parts come from this era.

We now know, of course, that electricity is fundamental to run electronic computers. But how? Back in the 17th and 18th centuries, electricity was one of those magical things dreamed up by crazy scientists.

The ancient Greeks came up with the word electron to describe amber, a yellowish, translucent stone that bugs were always getting stuck in. Turns out when you rubbed amber on fur, it would attract things, like feathers. If the Greeks had balloons and sweaters back then, they might have figured out electricity. In 1600, Dr. William Gilbert, the physician for Queen Elizabeth I and King James I, published a paper titled "De Magnete." Must have been for a French conference. On his days off from doctoring, he played with lodestone, and came up with the word electric to describe the magnetic-like properties of amber. He figured there were two types of electricity: Resinous, which sits around, and which the fur gave to the amber; and Vitreous, or fluid, which silk might give to glass when rubbed on it. Hey, progress, but still not useful for anything.

In 1747, Ben Franklin and an Englishman named William Watson independently figured out there was only one type of electricity, the vitreous kind, which like a fluid, flowed from one object to another. Franklin had a problem. The English language didn't have any words to describe the things he was pondering. So he invented whatever he needed: charge, discharge, battery, negative, minus, plus, conductor, electric shock (from his kite flying days in 1752). Franklin postulated that if an object had the fluid it was positive, and if it didn't, it was negative, and then decided that electricity flowed from positive to negative or from plus to minus. He had a 50-50 shot at getting this right, but blew it. He got it backward. As every schoolkid now knows, electrons are negative and flow, in your flashlight, from minus to plus. It would take the creation of semiconductors and the flow of holes in the second half of the 20th century, critical to how chips and computers work, to vindicate Franklin for his choice.

Priestley did some experiments in 1767 involving the measurement of electric force. Had he killed any frogs, he might have a more prominent place in the history of electricity. In 1780, Italian Luigi Galvani and his assistant were dissecting frogs, and the assistant dropped a metal scalpel on a frog's leg (yes, apparently they tasted like chicken back then as well). The frog twitched and Luigi proudly announced his discovery of "animal electricity." He then spent 20 years doing to frogs and birds just about anything he could think of including touching their legs and wings with brass and iron and copper and zinc, and hanging them on brass hooks.

Another Italian, University of Pavia Professor Alessandro Volta, was a skeptic of animal electricity. Instead, he was more intrigued by the metals. Volta figured the frog was just a detector of electricity, not the source. He visited Galvani and noted that when Galvani used a steel knife and a tin plate, the frog moved. So Volta put a frog between a stack of brass and iron and sure enough it twitched. But then he took copper and zinc and stacked them with a piece of wet pasteboard, and when he connected each end to a frog's leg, it practically jumped across the room. He realized that the electricity was created from a chemical reaction. In 1800, he alternated copper and zinc and silver, separated by salt water–soaked felt, and invented the battery. Today, appropriately, a Volt is the amount of electrical charge or amount of work the electrical charge can do and a Galvanometer is a measuring device.

While electricity would eventually be key to creating computers and power, no one bothered to make the connection quite yet. Frog's legs were one thing, but there was an Industrial Revolution going on, and more important tasks to be done.

Remember the Edmund Cartwright automated loom from 1785? Stream engines would drive the looms, but the cloth that came out was one color or one thickness. The bland Brits didn't complain, at least it was more comfortable than itchy wool. The French, on the other hand, demanded a little style, and were willing to pay for it. The son of a French silk weaver, Joseph-Marie Jacquard had a thriving business operating looms. But to meet the demands of discerning customers for interesting patterns, he needed weavers to lift or depress warp threads before each pass of the shuttle of the loom. This was painstaking work so output was slow and expensive. In 1801, Jacquard came up with an automated loom that operated with a set of punched cards. If there was a hole in the punch card, a spring-loaded pin and corresponding warp thread would be depressed. On the other hand, if there was no hole in the punch card, the pin would lift the warp thread. Jacquard even figured out how to create a loop of punched cards so patterns could repeat. A Jacquard loom was destroyed in the public square in Lyon in 1806. It didn't stop progress—by 1812, there were an amazing 18,000 Jacquard looms in France. Fashion anyone? Jacquard was awarded a lifetime pension by Napoleon and unlike anyone else in this story, Jacquard has a pattern named after him.

Jacquard looms made their way to England in the 1820s and by 1833, there were more than 100,000 working Power Looms. Surprise, surprise, not everyone was excited about this development. Disgruntled weavers in England burnt many a Jacquard loom. Others learned to shut them down by throwing a wooden shoe, known as a sabot in French, into the loom, and so became known as saboteurs. The Jacquard looms were the first mechanical computers used for commerce, as opposed to the Pascaline for finance. The memory was punch card with holes or no holes representing binary 1s and 0s. The logic was the spring-loaded pins that depressed or lifted the warp thread. OK, no one was surfing the Web with this computer, but Jacquard set up some basic computer concepts others would build on.

Punch cards? Hmm. Put that into our satchel—we might need that later, too.

Simple arithmetic and patterns for looms are one thing, but if mathematicians wanted to do anything more, they did it by hand. The Holy Grail for scientists at the time was to solve differential equations. Astronomers who studied the skies needed differential equations to predict orbits. The invention of the steam engine would have gone a lot faster if James Watt had been able to solve differentials in Isaac Newton's law of cooling. Newton stated that if an object, hot or cold, is in an area of constant temperature, then its rate of temperature change is proportional to the difference between the object's temperature and the ambient temperature. Newton figured this out empirically with a thermometer. Watt ran experiments to figure out how fast steam would cool.

In 1822, a flamboyant professor in England, Charles Babbage announced that he would build a Difference Engine. The size of a house, it would need a steam engine to operate but it would solve differential equations. Great idea, poor execution. A few small-scale models were demonstrated, but the engine was ahead of its time, by probably 100 years. After 10 years of trying, he gave up. He brooded around for a while and then in 1837 announced a somewhat less ambitious plan, the Analytic Engine, which would do math faster and to a larger scale than a Pascaline. Babbage complained that a Railroad Mania, then raging, hired away all the skilled workers and his Computing Engines were, at least partially, an attempt to do without them.

The daughter of poet Lord Byron, the lovely Augusta Ada King, Countess of Lovelace, was his assistant. This was key in getting government funding. Again, a few models were demonstrated, but like his Difference Engine, the Babbage Analytic Engine never actually worked. Still, he published many papers describing how the engine would operate if he built it. Much like the Jacquard loom, it had punch cards that contained the program and that would be fed into the Engine, which would run the program and spit out a result. This was the first idea for a stored-program computer but it would lie dormant until World War II. Babbage's son played around with models of it late in the century that actually computed pi to 29 places; a carriage jammed while computing the 30th.

The Analytic Engine, on paper anyway, was a decimal

machine—it did calculations using digits 0 through 9. This added enormous complexity to any calculating device, one of the reasons the Analytic Engine never went beyond concept. In 1847, at the ripe age of 32, teacher and mathematician George Boole wrote a paper, *The Mathematical Analysis of Logic.* This was controversial because logic, as Aristotle had defined it way back in 350 BC, was used exclusively to study all that touchy-feely California hot-tubby metaphysical stuff about being and soul. Important sure, but Boole suggested math could harness the same rules of logic. In 1854 he published *An Investigation of the Laws of Thought, on Which Are Founded the Mathematical Theories of Logic and Probabilities,* which laid out how to do math with binary logic, now known as Boolean Algebra, and which generations of engineers would later ace in seventh grade. This set up the incredible simplification of computers in the 20th century, which only had to deal with two digits, 1s and 0s, true and false, on and off, instead of 10 numbers.

This is a major inflection point in harnessing Logic and Memory for computers. Dealing with two instead of ten states lowered the complexity of computing devices by at least a factor of 10.

Meanwhile, the Industrial Revolution kept cranking. Amazing as it sounds, the British Empire really began with the invention of affordable, comfortable clothing. Silk was comfortable but painfully expensive, since there was no way to push productivity out of worms. Wool was cheaper, and warm, but way too itchy. We all know that.

It was the act of substitution that fired the economic engine. Cotton was itchy too, until it was stretched and wound into thread and yarn by the Crompton Spinning

Mule. This guy should get as much credit as James Watt. The Mule could only operate under power. Horses wouldn't do, a waterwheel might work, but a steam engine was exactly what was needed to give the mule enough power to stretch and wind, and make cotton "as smooth as silk."

British cotton and subsequent cloth made with automatic looms were of higher quality and lower cost than anything that could be done at home by old spinsters. This is exactly the kind of economic model that sparks massive growth. It's called elasticity—lower costs create new markets that never before existed. Everybody wanted comfortable clothing— once they could afford them. At a minimum, the world demanded comfortable undergarments, smooth against the skin, to shield against the itchy wool and scratchy burlap outer garments that were the rage. But you couldn't make smooth undergarments at home, even if you wanted to. You needed power and the Spinning Mule. As long as England could keep prices for its cloth down, it would both create new markets AND stave off competition from substitution. Growth protected, and created, by its own declining price elasticity. Put that in your pipe and smoke it. We'll see it again with semiconductors and integrated circuits in another couple of hundred years.

Transportation
Elasticity, Sea and Rail

Even in the early 1800s, the British Empire got off to a slow start.

The transatlantic slave trade was abolished in 1807, but it wasn't until 20 years later that the British would declare the slave trade a form of piracy, punishable by death. In 1833, slavery was abolished in the entire British Empire, after a five-year trial period, leading to the more commonly known date of abolishment in 1838.

The delay didn't hurt the engine. The plantations in the American South still had slaves, until 1865, and were cranking out cotton for the British textile mills. But there were still a lot of costs to be wrung out of the system, mainly in transportation costs, which could sometimes run three quarters of the price of English goods in foreign markets. Sail ships were just too small, slow and unreliable. That would soon change. The boost to elasticity was needed, and oddly, the inventions would come from the New World, where there was a pressing need for powered ships to move raw materials up and down American rivers to get to ports to load onto British ships.

John Fitch was the first to hook up a crude steam engine to paddle wheels. In 1787, he steamed a ship from Philadelphia to Burlington, New Jersey. Sure, this was only 20 miles, but it beat tacking a sailboat back and forth to get upstream. He received a patent for this steamboat in 1791, one of the first the new United States granted. But like Watt's original steam engine in 1770, Fitch's cylinder and pistons leaked badly, and Fitch didn't have John Wilkinson to come along with his cannon-barrel-boring tool. The leaky cylinder meant that Fitch, with probably a 3 horsepower engine, had economic problems, as he couldn't run cargo or passengers cheap enough and soon failed.

Back in England, James Watt wasn't resting on his laurels. His steam engine patent was to expire in 1800, so he kept inventing. As I noted, in 1782, he invented the double acting, noncondensing engine. Instead of just using a vacuum to "pull" down the piston, the double acting engine used steam to push the piston, first in one direction, and then in the other. The steam was never condensed; it was just expelled after it was used to push the piston. This was the realization of the Huygens engine, using force to move the piston, in this case expanding steam.

One guy playing with these things was William Symington. He had built an atmospheric steam engine for a steamship in 1788, but it was too much like Watt's and violated the patent. So he waited until 1801 to try again. This time, he built a low-pressure double acting engine, of his own design, but based on Watt's concept. In 1802, Symington successfully tested his steam-powered ship, the *Charlotte Dundas*, on the Forth-Clyde canal in Scotland. An American inventor named Robert Fulton happened to be there that day, and was intrigued. He moved to Paris to try to sell the French on his idea for a submarine, but they threw him out. Instead, he met Robert Livingston, then the American minister to France who was negotiating the

Louisiana Purchase. They soon became business partners and, in 1803, tested their own steamship on the Seine. It worked, but Fulton's engine was lame.

So in 1806 he went to the source, and actually bought a Boulton and Watt steam engine and had it shipped to New York. It was one of three steam engines to leave England, because until 1820, it was illegal to export steam engines from England without permission from the king. The other two, also made by Boulton and Watt, were pumping engines to supply water for the city of Paris and the city of New York.

In 1807, Fulton built the 142-foot-long *Clermont*, named for the New York town where Livingston was born, which soon made the 150-mile trip from New York City to Albany. It took 30 hours, including an overnight stop, which is what it feels like today on a Friday afternoon in rush hour. And it was commercially successful. Livingston and Fulton got a monopoly from the State of New York for steam navigation on the Hudson River. Their competitor, a guy named John Stevens, had a steamship named *Phoenix,* which, barred from the Hudson, became the first oceangoing steamship. Colonel Stevens hated the paddle wheels that he and Fulton were using, and toiled in his shop to create propellers driven by screw drives. He was off by about 40 years.

Livingston and Fulton had bigger plans. Livingston cut a deal with his old pals in the French government for an 18-year monopoly to run steamships in Orleans, part of their Louisiana Possession, at the mouth of the Mississippi River, knowing full well it would be purchased by the U.S. It's nice to have connections.

James Watt, meanwhile, was still haunted by the bad rap that the early high-pressure steam engines got when their boilers exploded, and he refused to use high-pressure steam. But others eventually would. High-pressure steam provided much more horsepower for the same displacement cylinders and the same weight engines. Watt's engine was all right for factories, even for paddleboats on a flat river, but pulling 30 tons of coal uphill on a railroad track would have required an impractically large, low-pressure Boulton and Watt steam engine.

Until 1800, the Watt patent ruled. Richard Trevithick, a bright engineer who worked with Watt's assistant, learned a lot about steam engines and eventually designed his own, using condensers similar to Watt's. He lost every legal battle that he and his partners waged to kill the Watt patent. So he had to find something else to do, and worked on creating a steam-powered carriage. In 1803, he showed off his carriage in London, driving it part of the way there, using wheels with spikes in them, kind of like golf shoes. Besides a few log tracks for horse-drawn trains, there were no rails or roads, just horse trails. And so in effect, Trevithick had invented the off-road vehicle.

After the demonstration, he took his steam carriage apart, sold it off and set out to create a railroad locomotive. By 1808 he had a high-pressure engine in a locomotive that operated on a circular rail line he had built in London, running up to 20 miles an hour with a piston over 2 feet in diameter and 6 feet long. The track eventually broke down and the locomotive jumped the track and crashed. Trevithick never rebuilt it, and moved onto rock drills and dredgers and tunnel diggers. He died penniless, and high-pressure engines were still not perfected, but he proved that construction materials and techniques had advanced enough to allow for safe usage of high-pressure steam.

Others saw his work and kept it going, the most important was George Stephenson. In 1803, he was working in a mining pit, and was put in charge of repairing an old New-comen steam engine, still in use, probably because the mine was too poor to upgrade. There is something oddly magical about these Newcomen engines—as bad as they were, peo-ple who touched them went on to big things.

In 1815, Lord Ravensworth, owner of the Killingworth colliery (that's coal mine to you and me), advanced Stephen-son money so that he could create a steam locomotive to haul materials out of the mine. His first attempt, the *Blutcher,* was a double acting engine, like Watt's, but didn't connect directly to the wheels. It worked but was no cheaper than horses to operate. Instead of cogs and pinions and spikes, the *Blutcher* worked on tracks, with flanged wheels sticking to the parallel railroad tracks. Friction works.

His second engine fixed all the shortcomings of the *Blutcher.* It was a high-pressure reciprocating engine, with two cylinders that pumped in opposite directions, the push me, pull you of Dr. Doolittle fame. Rods were con-nected directly from the pistons to the wheels, one side being a quarter of a turn ahead of the other so the engine never got stuck straightforward or straight back. A bit like bicycle pedals—one piston was always pushing a wheel forward.

Coal was burned in a firebox, which sat at the back of a huge boiler that created the high-pressure steam. After the steam pumped the piston, it was expelled out a smokestack that was also the chimney of the firebox, helping create a draft to stoke the fire. Pretty ingenious. This also explains how the Little Engine That Could could go Puff Puff Chug Chug (over the mountain to bring toys to all the little boys and girls).

Over the next five years, he built 16 locomotives, each better than the last. In 1819, he was asked to build railroad lines. In 1821, Parliament authorized a horse-drawn 12-mile rail line between the coalfields in Darlington to the river ports in Stockton. Edward Pearse, the major shareholder of the Stockton and Darlington Railway, met with George Stephenson, who told him to forget the horsies. His *Blutcher* locomotive could do the work of 50 horses. A quick showing in Killingworth of the *Blutcher* in action clinched the deal.

George then went into business with his son Robert. They established Robert Stephenson and Co., with 1,000 pounds of capital from Pearse, who knew a great investment when he saw one. This turned out to be a fruitful venture all around, as it helped complete the Stockton and Darlington line in 1825, with locomotives that could haul 450 passengers at 15 miles per hour. Stephenson insisted on malleable wrought iron in 15-foot lengths from the Bedlington Ironworks for the rail line.

Of course, Stephenson created a virtual circle. The Killingworth coal mine and the new Stockton–Darlington line helped to transport coal to the coke ovens of ironworks, so the company could produce better rail lines and deliver more coal. This is similar to Silicon Valley today, where companies race to produce faster computers that engineers use to design faster chips, so that companies can produce faster computers. It's a tight circle, but those outside of it benefit from the improvements.

In 1826, Robert Stephenson and Co. was asked to help design the 36-mile Liverpool–Manchester line. This was the missing link of the Industrial Revolution. Manchester was the key industrial center of England, laden with ironworks and mills and manufactories. Liverpool was the closest seaport. A railroad running between them was no longer an option; it was mandatory to both lower transportation costs and increase volumes.

To secure the final contract with the Liverpool–Manchester line, it had to prove its locomotives in a speed trial of locomotives, with a minimum speed cutoff of 10 miles per hour. This was the drag race of the time, but with a real purpose, as many still believed that horses should pull trains on the line.

Four engines entered the contest at Rainhill for the 500 pound prize and perhaps the rights to build locomotives for the line. They were *Novelty, Sans Pareil* (literally translated: without similar), *Perseverance* and Stephenson's new locomotive aptly named *Rocket*. A fifth entry named *Cycloped* was powered by a horse running on a belt—think of a horse on a treadmill—but was withdrawn when the horse fell through the floor.

Next out of the contest was the *Sans Pareil*, which was a fuel hog, and a design flaw gave it a huge blast that caused unspent coal to be blown out its vent, an ugly sight. It suffered a cracked cylinder and its quick exit was strange since the company that manufactured the cylinder was owned by George Stephenson. Hmm. *Perseverance* couldn't get to 10 miles per hour and was scratched. That left *Novelty* and *Rocket*.

In the first run, Stephenson's *Rocket* clocked in at around 12–13 miles per hour. Not bad. The *Novelty* then ran at 28 miles per hour. The *Novelty* had a decent engine, but instead of a vent providing draft for the furnace, it had a set of bellows to stoke the flames. Before the end of the first day, the bellows broke down, its joints had busted and the *Novelty* was sidelined.

That left the slower but dependable *Rocket*. Its design was unique. Stephenson had increased the steam-generating capabilities by replacing a 12-inch tube in the boiler with a series of 2-inch copper tubes, thereby increasing the surface area heated. The vertical cylinders were changed to a 35-degree angle, which provided more power to the wheels. On the second day, the Stephensons just turned up the steam pressure and *Rocket* ran at 25 miles per hour, then 30, even briefly hitting a top speed of 36 miles per hour. It won the contest and the Stephenson Company won the contract for the Liverpool–Manchester line.

On June 14, 1830, a new locomotive similar to *Rocket* named *Arrow* ran on the completed line. It was predicted that 400 passengers a day would ride the line, but soon the number rose to 1,200. By 1835, the line had carried half a million passengers.

Demand for railroads, from passengers and for industrial goods, exploded. You could put in a 20-mile railroad for the equivalent of $650,000 and collect that much in fees every year. Joint stock companies were the rage, and the stock market was all too happy to step in and provide capital. And more capital. And then too much capital.

By the 1840s, a "Railroad Mania" was raging, with stocks selling on multiples of passenger miles, a precursor for multiples of page views that Yahoo! stock would trade on 150 years later. Charles Babbage complained that "the railroad mania withdrew from other pursuits the most intellectual and skillful draftsmen." He sought to invent a machine that might replace them, and although he couldn't have foreseen it, make Yahoo! possible. This is when Charles Dickens marveled at railroad wealth. Investors made money, investors lost money, but in the best and worst of times, the railroads got built, and people and goods were shuffled about.

The Industrial Revolution had hit its stride.

• • •

Around the same time, these high-pressure steam engines were strong enough and reliable enough for oceangoing vessels. Even then, paddle wheels powered the first oceangoing steamships.

In 1819, the *Robert Fulton,* a steamship designed for ocean travel, went from New York to Havana, one assumes to pick up cigars. Most ships crossing the Atlantic were converted to combination sail and steam ships, the steam for emergencies or periods of no wind. One such ship, the *Savannah,* with side paddles, steamed out of Savannah, Georgia, on May 22, 1819, cleared land, hoisted its sails, and rode the winds to Liverpool, where 29 days later it dropped its sails and steamed into port.

One person there that day was an American named Junius Smith, and he began to dream about a steam-only trip across the Atlantic. There was only one problem, how to carry enough coal to keep the steam engine cranking for that long trip. A self-proclaimed "expert" on the subject, the Reverend Dionysius Lardner, proclaimed in 1837 that the longest theoretical distance a steamship that carried its own coal could travel was 2,500 miles. He basically made up the number!

Smith figured he knew math better than the great reverend. The volume of a ship, and therefore how much fuel it could carry, goes up by the cube of the ship's length. But the amount of fuel a ship requires is in relation to driving the bottom surface of the ship through the water. And surface area only goes up by the square of its length. If you could build a long enough ship, you could travel the globe.

Smith heard about an engineer named Isambard Kingdom Brunel, who had made a fortune with his Great Western Railway, in the wake of the Stephensons' success. Brunel also figured the Reverend was blowing hot air, and in 1837, he began construction on a monstrous 236-foot steamship named the *Great Western*.

Junius Smith freaked. He had set up a company to build a similar ship, but now he had to act fast. He bought a dumpy coastal steamship, christened it the *Sirius* and retrofitted it for an Atlantic crossing. In March of 1838, the *Sirius* took off for New York. Now Isambard freaked, and three days after Smith, he launched the *Great Western* for New York to try to be the first to cross the Atlantic under steam power.

The race was on.

Smith, whose ship wasn't as big as the *Great Western*, miscalculated on how much coal he would need, and a strong headwind made matters worse. He ran out of coal before reaching New York, and threw anything that would burn into the furnace. He reached Long Island on April 22 and anchored in New York Harbor on April 23. Eight hours later, I. K. Brunel and the *Great Western* pulled up and anchored alongside, her coal bins still one-quarter full. Transportation costs were set to drop, as a steamship could carry 10 times or more cargo than a sail ship. This cost reduction would take another 10 years to play out, but was just the cost reduction/volume expansion that would keep the British-centric Industrial Revolution running for another 60 years.

Sirius and *Great Western* were both driven by paddle wheels. Like me, you are probably imagining one of those quaint Mississippi paddle wheel steamers, heading up the lazy river with Mark Twain onboard smoking a corncob pipe. You wouldn't be that far off, but it is not the kind of ship you would want battling 30-foot rollers on the high seas. Plus, they were painfully slow.

Colonel Stevens had tried in 1807 to use screw-driven propellers instead of paddle wheels. He ran into all sorts of problems, the main one being the inability to get a machine shop to build strong enough propellers or shafts of the right tolerance. So he gave up and just copied Watt's steam engine and Fulton's paddle wheels.

It would take another 40 years for screw propellers to become the rage.

In the summer of 1839, the steamship *Archimedes*, equipped with screw propellers, arrived in London after traveling along the coast of England. Archimedes of Syracuse (Greece, not New York) had invented the screw pump to raise water for irrigation back in 220 or so BC. The steam engine would directly drive a shaft to which a propeller was attached. The screw propeller was more efficient than a paddle wheel, because the moving water running past the ship, in other words the wake you might water-ski on, actually helped turn the screws, the opposite of the way that Archimedes figured it would be used. So once the ship was in motion, it took less power to keep it in motion. The *Archimedes* could run at 10 knots, and used half as much fuel as a paddle wheel ship.

Now the race was on to run the first screw propeller steamship across the Atlantic. Our old friend Brunel scrambled to take that honor. But this time, he would do it in style. The profits from running the *Great Western* across the Atlantic gave him the funds to build the *Great Britain*, a 3,600-ton iron-hulled beast capable of holding 252 paying passengers and a crew of 130.

It was launched with great fanfare by no less than Prince Albert in July of 1843, but was not yet seaworthy, and was repaired in a floating dock and launched again in December of 1844. It took off for New York on July 26, 1845, arriving 14 days later, and Brunel took the honor of the first steam-driven propeller crossing.

At 322 feet long and 51 feet wide, the *Great Britain* must have been some sight pulling into New York Harbor. It contained four 6-foot diameter cylinders, together capable of 1,600 horsepower, which drove a single 4-ton, 6-blade propeller measuring 15½ feet in diameter. The boilers were fed 70 tons of coal daily. The cylinders turned 18 times per minute, but gearing got the propeller to turn 53 rotations per minute, enough to get the *Great Britain* to run more than 11 knots per hour (12 mph). Remarkably, the engines of the *Great Britain* still used low-pressure steam, a mere five pounds per square inch.

Brunel had even bigger dreams. He built the *Great Eastern,* a 688-foot long paddle and propeller steamship capable of holding 4,000 passengers. It was launched in 1858 but didn't cross the Atlantic until 1860, with only 35 paying customers.

It was a financial failure. The opening of the amazing Suez Canal did away with many of the long voyages that Brunel envisioned. It was sold for a song in 1864, but ended up with a place in history. The Telegraph Construction and Maintenance Company contracted with the *Great Eastern*'s new owners, the Great Eastern Steamship Company, to lay cable that would connect England to North America in 1866. You see, it was the only ship large enough to hold the massive spools of telegraph cable to be laid on the ocean floor. But I'll get to that later.

The Suez Canal was the next cost reducer. With the Atlantic tamed by steam, the next challenge to lower transportation costs was a route to India and China. Going around Africa's Cape of Good Hope (the name says it all) was long and treacherous. It was worse for steamships, because they needed coal along the route. Many ships traversed the Mediterranean and dropped their cargo in Egypt, where it was transported to the Red Sea and reloaded on other ships for the rest of the trip. You can imagine all the goods that "fell off the loading dock" into locals hands. Napoleon considered building a canal back in 1798, but was told that there was a 9-foot difference in sea levels. Plus, while he was thinking it over, Admiral Horatio Nelson sunk half the Emperor's fleet at the Bay of Aboukir.

In 1854, the French engineer Ferdinand De Lesseps cut a deal with the Egyptian government, the First Concession, for the rights to dig a canal between the Med and the Red. Did his parents know when they named him Ferdinand that he would shorten the route around the world that Ferdinand Magellan started in 1519? He created a company called the Compagnie Universelle du Canal de Suez. Couldn't they have just called the company Canalco? Suez was the name of the Red Sea port to be connected to the Mediterranean Port Said 40 miles away.

Starting in 1859, some two and a half million Egyptians were conscripted into slavery to dig out the canal and 125,000 of them died. There arose no Moses to throw a few plagues around and declare "Let my people go." De Lesseps ran out of money, and the head of Egypt, Pasha Said, stepped in and bought 44% of the company. Then in 1863, the British woke up and realized that the French were building a canal that would have serious strategic implications. So they and the Turks managed to put a stop to the digging. Napoleon III helped form an international organization to oversee the canal, and work started again in 1864. The two sea levels were so close that no locks were needed (to raise or lower ships to different water levels).

The Suez Canal was first opened to traffic on November 17, 1869. The implications to world trade were amazing. Transportation costs dropped, depending on distance, by a factor of three or more. As important, distance and time became deterministic. The trip from Malaya to England to deliver tin took exactly three months, which was the same time it took for copper to arrive from Chile. This allowed commodity exchanges, like the London Metal Exchange, to create three-month forward contracts. Contracts for the purchase or sale of a commodity three months into the future allowed buyers or shippers to hedge their business, lowering the cost of risk. We'll need these financial markets in this story soon.

Within three years, the canal almost went bankrupt, and Egypt was forced to buy most of the shares of the Suez Canal Compagnie it didn't own, mortgaging the future profits of the only decent business it had: cotton (who can resist the subtle feel of Egyptian weave bath towels?).

By 1875, Egypt was going broke. Who needed to hire local workers to load or unload when you could just cruise through the canal? Cotton profits were going to the bank. In November of 1875, the British government bought the Egyptian shares of Compagnie Universelle du Canal de Suez for 17 million pounds, and sent in troops to take control of the canal.

Of course they did, as the benefit to their economic engine cannot be overstated. Their empire was global and sea trade was the glue that tied it all together. Delivery times and therefore costs just dropped by such a huge amount, 50–80%, that the Industrial Revolution, which was effectively over, got an end-of-cycle cost kicker that would sustain it for another 40 years.

It is odd that the French, and a guy named Napoleon (III) deserve credit for this. De Lesseps became a world-renowned superstar but would later meet his match trying to build the Panama Canal.

Machine tooling had advanced since Wilkinson's days, to the point that tight boilers could be made to increase steam pressure with somewhat lower risk of explosions.

Once high-pressure steam was safe, work began on more efficient ways of using it. The first real improvement was compounding. Rather than just crank up the pressure of steam fed into cylinders, and then expelling it after it pushed the piston, the still high-pressure steam could be fed into another cylinder. It would then push another piston, and then be expelled. Triple and quadruple expansion engines just add more cylinders so that the initial high-pressure steam was not wasted. This allowed steamships to go twice as fast on the same amount of fuel as the 11-knot *Great Britain*, and cut the time of all voyages. By 1886, the transatlantic journey would take only seven days.

In 1884, an Englishman named Charles Parsons would take the next big leap. He studied a failed design by a Swedish scientist named de Laval for a set of rotating blades, spun at high speeds by high-pressure steam coming out of a flared nozzle. Parsons fixed the design by having the steam expand in multiple stages, and move through an alternating series of fixed and rotating blades. His invention was the impulse turbine engine. By getting it to spin at lower speeds, he made it practical, and promptly patented his turbine.

In 1893, Parsons and five associates founded the Marine Steam Turbine Company, and in 1894, they launched HMS *Turbinia*. Three turbines were set up as a compound engine, meaning the exhaust of one drove the next one. This created a combined 2,000 horsepower, which drove three shafts connected to three-blade propellers. They spun at 2,500 rotations per minute, or 50 times faster than the monster propeller on the *Great Britain*. It was really just a demonstration ship, created to help sell turbines to the Royal Navy and the big cruise ships. The 104-foot-long, 9-foot-wide ship was sleek and could go 34 knots (40 miles per hour).

At Queen Victoria's Diamond Jubilee, the *Turbinia* showed off by cutting in and out of the progression of the (slow) Royal Navy ships of the line. By 1905, the Allan Line's *Virginian* and *Victorian*, and the Cunard Line's *Mauretania*, *Lusitania*, and *Aquitania* all had Parsons turbines. In 1906, the HMS *Dreadnought* warship launched. It was 526 feet long, and its four Parsons steam turbines provided almost 25,000 horsepower.

By the way, the reciprocating cylinder engines didn't go away, and coexisted with Parsons' turbine. One subtle feature of the turbine is that it does not go in reverse. So while these big, oceanliners were out at sea, they were under the more efficient turbine power, and when maneuvering around ports, the cylinder engines took over so that the propellers could go in reverse.

The unsinkable *Titanic* was fitted with both engine types. The turbines were Parsons'. When the fateful iceberg was spotted, the 37 seconds until collision was not enough time to switch over to the cylinder engines/maneuver mode to either turn or go in reverse.

Parsons' turbines had other uses beyond marine engines and indirectly helped push the modern computer business into being. Bookbinder-turned-physics-wizard Michael Faraday showed that when wires were rotated next to magnets, they created electricity. Turbines, driven by steam engines or flowing water, would be the perfect solution for generating cheap electricity in the 20th century.

What a long way the steam engine had come. Watt improved the Newcomen design and improved its power from 3–5 horsepower to maybe 25 horsepower. By the end of the 19th century, steam power had increased by a factor of 1,000; on a warship to protect trade routes, those very same steam engines enabled trade.

I think the lesson here is not any specific piece for computing, but instead the parallels of the Industrial Revolution and the digital revolution. Elasticity and scale, or the ability to constantly lower the price of goods and services, drove the industrial era. So too, there are parallels of the microprocessor and networking with steam engines and transportation. They go hand in hand.

What else? Well, these are 100-year cycles. And it's really important to have a thriving domestic market. The French didn't invent railroads because they didn't need to get coal out of coal mines, or get raw materials and finished goods between ports and factories. Same with turbines, if you are shipping goods worldwide, your own inventors and innovators look for solutions. You have to know what the problem is to even begin to look for the solution.

One thing missing in all this is how world trade took place. What was used for money, how was it all funded? Let's see.

PART 2:

EARLY CAPITAL MARKETS

Funding British Trade

Capital markets were hungry to fund British Triangle Trade expeditions—slaves for raw materials like cotton and tobacco, and lumber for finished goods. As long as no one sunk the merchant ships, the expeditions were almost guaranteed to be fantastically profitable. Despite the emergence of joint stock companies, many expeditions were funded by bankers and insured by underwriters working out of coffee houses. Since the British government was collecting custom duties on tobacco and on cotton, it made sure to fund the British Navy to protect these lucrative trips. We need to go back a little, and find out how this navy, and the capital markets, came about.

It was King Henry the VIIIth (I am) who got England out to sea. His dad had built a flotilla of merchant ships, but H8 commanded the lamest of navies to protect them, five barely seaworthy dinghies. That is, until the royal designers mounted cannons on the side of the *Mary Rose,* which launched in 1511. And that innovation made all the difference in naval warfare. The Brits now had the ability to fire broadside, a nasty change in sea wars. With a new class of lethal sea power, Henry grew the navy to 53 ships, but never did much with it, and the Spanish and French soon copied the side-mounted cannon design.

Henry was more obsessed with creating Henry the IXth, but his wife, Catherine of Aragon produced only a daughter, Mary. Divorce was out of the question, at least with the Catholic Church, so as quick as you could say "off with her head," in 1527 he initiated the English Protestant Reformation. As Supreme Head of the Church of England he could not only sign his own divorce papers, but whack his own chancellor Sir Thomas More for arguing with his promiscuity. Two plus centuries later, Henry's hankerings would play a big role in the Industrial Revolution, but we'll get to that. After old Henry died from six hen-peckings, Catholicism made a small comeback. His daughter Mary married staunch Catholic Philip II of Spain and is forever known as Bloody Mary for torching 300 Protestants at the stake in the name of the Church. As you can imagine, Protestantism came back by popular demand, and Henry's daughter by second wife Anne Boleyn became Queen Elizabeth I in 1558. Think of her when you trade stocks.

The competition at sea was with the Spanish and the Portuguese, who were both squeezed from the French in the north and Muslims to the south. So they headed west to the New World and east to the real India. By the time Liz took office, Spain had pillaged the Aztecs of their gold and the rest of the Americas of their silver, and was shipping it back to pay folks in the East Indies for spices. This would not be the last time that a rush for gold led to increased trade and booming economic times.

Elizabeth the One kept daddy's religious convictions, and hated the Catholic nations. But instead of declaring war and attacking them directly, she sanctioned piracy against Spain's ships. Sir Francis Drake considered himself a privateer, but was more of a racketeer, paying protection money back to the Queen. Drake and his cousin John Hawkins (was everyone back then related?) raided the Spanish and Portuguese ships in the routes to the East Indies, West Indies and anywhere else they could pillage. Sir Walter Raleigh earned his title as a privateer.

Meanwhile, Elizabeth's cousin Mary Queen of Scots up north had ambitions to be Queen of England, even marrying and then murdering her cousin to put herself in line to the throne. Three years later, Protestant landowners in Scotland tossed her out, so she sought asylum in England but Elizabeth had her jailed. To remove any potential threats to her throne, in 1587 she had her cousin beheaded, but not before Mary gave birth to James, soon to be known as James I.

Drake, meanwhile, was a bit of a pyromaniac. In 1586, he lit up a number of Spanish ships in the harbor of Cadiz, in an act known as "singeing the King of Spain's beard."

The Spanish could only take so much of this. In 1588, King Philip II sent 150 ships, the famous Spanish Armada, to land a force on English soil, avenge Mary Queen of Scots' beheading, reinstate Catholicism, and teach the English a lesson about pillaging. The Spaniards had lots of luck, all bad. A spot of bad weather, as the English are used to almost year round, blew them off course, and allowed the English to attack them from the rear. Drake even lit up eight old ships and had the wind push them into the Armada. In the first signs of English aptitude for war at sea, they sunk all but 65 ships of the Spanish Armada.

This cleared the way for the British to rule the sugar and slave trade in the West Indies and have a fighting chance for their share of the spice trade in the East Indies. Trading in commodities was worth more as an economic engine than pillaging gold, which merely glitters, a lesson still forgotten.

Elizabeth I understood this. She helped Sir Thomas Gresham set up the Royal Exchange in 1566, emulating a successful Dutch exchange. It mainly traded metals like tin and copper, but soon shares of companies like the Russia Company and Levant Company began to trade. These trading companies were provided access to capital to build ships and finance trading expeditions, and the era of organized capital markets for the English was born.

On December 31, 1600, when everyone else was singing "Auld Lang Syne," 218 knights and merchants in the city of London created the East Indies Company, and were given exclusive rights by the crown to all trade in the East Indies. At the time, this was no lay-up as the Dutch and those nasty Portuguese controlled the trade routes. But the EIC, with a trading business and its own private military to protect it, became the largest nongovernment organization. Think Exxon with weapons.

The creation of the East Indies Company was the first of many government-anointed trading companies that formed the backbone of mercantilism. Free enterprise was not yet ready for prime time; the monarchy was having too much fun doling out favors to friends of the crown. It worked for a while. There was so little trade with India that giving out a monopoly was no big deal. The next 200 years would see the growth of this mercantile system until Adam Smith hallucinated the Invisible Hand.

In the meantime, it turned out that Elizabeth I was way ahead of her time. She figured out that world trade and a strong navy were interdependent; she couldn't afford the navy without the wealth created by trade, and a capital markets system to fund it.

England had almost no colonies in the New World while Spain had spent the past 100 years exploring and controlling Central and South America and what is now Florida, Texas and California, giving it not only plenty of gold, but enough electoral votes to win any election. The defeat of the Spanish Armada provided Elizabeth an opening to compete with Spain but she was reluctant to provide funds, since John Cabot in 1497 and then Sir Walter Raleigh in 1584 had each taken a financial bath trying to create colonies.

In 1600, while the East Indies Company was being launched, England's cities were filling up with people from the countryside looking for work. New farming techniques had reduced the need for labor on farms. This would play out again 120 years later as good weather would lower crop prices sending people to cities looking for new ways of making a living.

A geographer named Richard Hakluyt (HAK-loot) suggested to Elizabeth in 1607 that she could ease the overcrowding and create colonies at the same time by shipping all the people streaming into London over to the New World. Elizabeth was intrigued, but resisted committing funds from her Treasury to fund it. But she did agree that others might want to fund these colonies.

The Royal Exchange was up and running and needed stocks to trade. Hakluyt and others suggested that funds could be raised from England's wealthy class. Elizabeth agreed and the joint stock company was born. Rather than be funded by the Crown or just one individual, capital instead would be raised from a large group of wealthy individuals, minimizing the risk for each. Elizabeth provided the license, so to speak, but the markets, which were nothing more than the pooled wealth of Elizabeth's wealthy subjects, provided the capital. Liquidity would be provided as shares of these joint stock companies traded hands on the Royal Exchange, or in private transactions out on "the Street."

In 1607, Hakluyt was still pitching his ideas for a fund to

cover the costs of creating colonies, this time to James I, son of Mary Queen of Scots, who had taken over the monarchy after the death of Elizabeth. In his "Reasons for Raising a Fund to Settle America On the Value of Colonies to England" (you can find the whole thing on the University of Chicago web site), he wrote:

> REASONS OR MOTIVES for the raising of a public stock to be employed for the peopling and discovering of such countries as may be found most convenient for the supply of those defects which this Realm of England . . .

Almost 400 years ago, Hakluyt laid out the reasons behind today's modern publicly traded stock, as well as its benefits. These include reduced risk for investors, personal incentives for owners, payoffs for the government in the form of duties and taxes, and benefits to the entire state. Think of this when you buy 100 shares of Amalgamated Mogul in your E*Trade account. What Hakluyt couldn't have possibly imagined, but alluded to, was that the stock market would be the great allocator of capital to these joint stock companies in such a way to constantly propel society forward on a vector of progress that no king could do on his own. It's odd that a tool of mercantilism, the ability to raise risk capital on the expectation of returns by selling shares in a company, is today the backbone of capitalism.

Joint stock companies were an interesting turning point for England. In Spain and France, the monarchies owned everything. In England, joint stock companies were almost a form of stock options. They extended property rights to individuals, AND provided a big fat carrot for the company to succeed. Individuals were empowered to create value, not for God and Country, but for themselves, which was Hakluyt's point. What a concept. But all was not rosy, at first.

The New World was quickly being settled and exploited. In 1607, King James I gave the Virginia Company a monopoly to exploit Virginia and for that received a fee, and also took a little piece of the action, owning a minority interest. Mercantilism had its benefits. But despite efforts to grow olives and export lumber, the Virginia Company went belly-up by 1624. James didn't last much longer and a year later, Charles I inherited his dad's throne.

In 1624, Parliament enacted an important piece of legislation. The English Monarchy helped fund the Parliamentarians' elaborate lifestyle by not only taxes and custom duties, but by selling off rights for monopolies in every known business, like the Virginia Company in the New World and the East India Company in India. You can imagine that patronage and favoritism were rampant. Parliament was filled with landowners, who often saw monopolies on trade increase their costs of doing business. Plus, they were on a kick to remove powers from the King. So they enacted the Statute of Monopolies, also known as the Monopoly Act, also known as the Patent Act. The King could no longer grant monopolies, but as only politicians can do, Parliament included some exceptions.

> Provided nevertheless . . . that any declaration before mentioned shall not extend to any letters patent or grants of privilege for the term of one and twenty years . . . to the first and true inventor or inventors of such manufactures which others at the time of the making of such letters patent or grants did not use. . . .

The Venetians had written a patent law in the 15th century, but it probably only protected window blinds. Parliament's tiny exception in 1624 set forth the greatest protection of intellectual property rights for writers and inventors. Well, limited protection, one to 20 years, but there are always exceptions to the exceptions. The "protection under law" of ideas prompted individualism, self-interested folks who could work hard knowing they could reap the benefit of their own work. Adam Smith would note this much later, but for now, it set off a wave of invention.

A strong and liquid capital market became an important component to enable the British Empire. They almost didn't have one.

Capital Markets and Bubbles

Today, money sloshes around the globe quite easily, from Zanzibar to Berkeley Square in milliseconds. But back in the 18th century, money was a local instrument. Almost by necessity, precious metals such as gold and silver were the de facto currency for trade—no one trusted much else. Monarchies and their governments created their own currencies, backed by gold, first as a convenience—a titan of industry would need wheelbarrows filled with gold to do his business. But soon, currencies became a tool to control the economy.

It is difficult for anyone to properly issue just the right amount of money to grow the economy, and the best structure was to set up a bank at arm's length to handle this function. As we'll see in a second, the British smartly set up the Bank of England, which sometimes competed with other banks, but would usually complement them by lending what was needed.

But it was stock markets that proved over time the mechanism to provide capital for growth. Exchanges have been traced back to the 12th and 13th century in Italian cities like Genoa, Milan and Florence. As world trade expanded, the Venetian and Florentine merchants would meet Norwegians in the Belgian town of Bruges to swap their currencies. They would meet in front of the home of the Van de Beurses family and trade during what became known as "de beurse" meetings. Not good enough to be invited inside, in 1531, they moved to Antwerp and opened their own building and set up the first of many European bourses, or exchanges. Banks and stock markets would compete for almost 500 years. Which are more important today?

In 17th century England, the ever-increasing power of Parliament established many important rights for the King's subjects, although I suppose at some point they actually became known as citizens. Since many in Parliament, especially the Whigs, were property owners, one of the key rights they pushed for was property rights.

The effect the Bill of Rights in 1689 had on the rule of law after the Glorious Revolution cannot be overstated. Although it didn't explicitly say so, this "Declaration of Rights" granted the people strong property rights. This meant you could own land or a business without the fear of some whacked-out, thinned-down bloodline king or vengeful politician taking it away from you. It would take an entire Parliament of vengeful politicians to do so, which of course was never out of the question.

Property rights were key in establishing meaningful businesses, and it also solidified the banking business and the business of extending credit, because now an institution could lend you money with your land or business as collateral. It could seize your land, but only if you screwed up bigtime. But just the fact that the king couldn't take your property away opened up the possibility of capital markets.

The Monarchy was no longer wealthy and credit was tight. It only had so much gold, and couldn't just issue money willy-nilly.

A Scottish financier by the name of William Paterson had made a killing in trade. In 1691, he offered to loan the government 1.2 million pounds. What he asked for in exchange was a small concession, although one much bigger than the popcorn concession at the circus. He wanted to be the government's exclusive banker. As a result, in 1694, a Royal Charter created the Bank of England, a joint stock company with secret owners. I don't know why it was so secret, probably a few too many "friends of the crown."

In 1707, under Queen Anne, England and Scotland merged in a nonhostile takeover, creating Great Britain. At the time, that was a rather presumptuous title, it didn't earn the name Great for another 91 years. But it was the largest free trade area of its time, and this helped to create enough wealth to fund a stronger navy to protect merchant ships trading with the New World.

Believe it or not, England, and many other countries, had been raising money by selling life insurance and lottery tickets. For the first time, the government could borrow against its future taxes and custom duties, with a publicly traded security, and therefore fund bigger projects. As an aside an Englishman named Holland created the Bank of Scotland a year later, try to figure that one out.

But it wasn't just wealth that made the British economy so robust. The French had plenty of wealth, but the monarchy and government couldn't really get at it, not on a long-term basis anyway. Borrowing in France was money market borrowing, short-term stuff that had to be paid back quickly, in a year or less. Worse, the French borrowings were often via a bureaucratic system of offices. Those with wealth bought an office, a government job basically, and were paid a salary year in and year out. Anyone lending money to the government of France expected interest and a job!

The Bank of England coordinated debt financing, much as the Federal Reserve does in the U.S. today, and syndicated out loans to other banks. In effect, this monetized the government's debts, allowing them to be traded on a market and providing debt holders liquidity, meaning they could get rid of it quickly. This relief of not having to be stuck with government debt created a liquid market that enabled the government to borrow capital over long periods of time—decades instead of 12 months.

The ability to trade in and out of holdings is the key to capital markets. Investors are more interested in buying a security that they can sell, even if it means paying a premium for it, and even if they never do sell it. Value comes from the liquidity, or the ability to sell on demand. Until then, wealth was held in land and animals and equipment that were, and still are, hard to dispose of.

Spain had lots of gold, but not much else. The Spanish were terrible at trade, their navy subpar, and their global influence waning. In the early 18th century, England was at war with Spain over many of its colonies in the New World.

In 1711, Robert Hartley founded the South Sea Company. Actually, it was started by an Act of Parliament, which gave him an exclusive contract for all trade with Spanish possessions in the Caribbean and South America. But there was a twist. The South Sea Company was one of these new-fangled joint-stock companies, meaning it was a corporation with shares that could be transferred, meaning they were liquid. The way Hartley got his Act of Parliament was to agree to take on 9 million pounds of England's national debt.

Holders of that debt were forced to exchange it for shares of the South Sea Company at par, meaning a hundred pounds of debt was exchanged for a hundred pounds of stock. Value for stocks and bonds usually started at 100, or par. This was typically the net worth of the company and where it was originally valued. Like bonds, stocks would trade above or below par, depending on how they were doing or what interest rate they were paying. South Sea could then borrow against future interest payments, of around 6%, from the government. England ditched some of its debts, the South Sea Company could raise money by borrowing against future interest, and shareholders gave up their dividend for the upside of profits from trade with Spanish possessions. Hey, win-win all around.

Well, not so fast. A sure thing with a speculative upside is a precarious trade. No one knew the true value of the upside. A deal was reached with Spain in 1713, called the Peace of Utrecht. The Spanish kept most of their trading rights, South Sea got very little. Ouch.

So the company scrounged around for a new business. The only thing it was good at was making government debt disappear. Oh, and it had another asset, Robert Hartley, who was a phenomenal promoter. He made sure that shares of South Sea were in the hands of members of Parliament and newspaper publishers.

A director named John Blunt, who had written the company's original charter, took over. By 1719, Blunt noted that Parliament was struggling under about 30 million pounds of debt. Blunt offered to exchange debt for shares again. But this time, it would not be at par or its original value or true net worth, but at the value that South Sea shares were trading. This made it ripe for manipulation. In fact, the only true profit motive for the South Sea Company would soon become driving its stock price up over par and then exchanging it for debt, the profit being the difference between the stock's value and par. These profits would then be "reported" to investors, which would drive the stock higher. Circular and dangerously wrong reasoning. Enron would pull this scam almost 300 years later.

England was ripe for manipulation. At the exact same time, thanks to a Brit named John Law, the French were in the midst of their own speculative bubble: The Mississippi Company was worth more than all of the gold and silver in France. The Dutch tulip mania of the 1630s didn't quite kill off speculation there. Between the fall of 1719 and the summer of 1720, close to two hundred joint stock companies, many of them purely speculative, went public.

As an inducement to Parliament, Blunt offered a fee of 7.5 million pounds back to the Treasury for the rights to do the exchange. In early 1720, Parliament agreed to Blunt's swap proposal. It didn't hurt that he paid off the secretary of the treasury and the Duchess of Kendal, with whom King George I was doing his own speculating. In order to be able to afford the 7.5 million pounds and create profits for the company in doing the exchange, the South Sea stock would have to be a lot higher. It wasn't hard to get the stock up, it was thinly traded and buying a few shares could drive it up 50 points. The company sold over 2 million pounds of stock with the price around 300 pounds for each share, well over the 100 par price. They declared a 10% dividend, which got the shares to 340 pounds each.

You could buy shares of South Sea on an installment plan: Buy 100 shares, but only pay for 20 up front. Then the company announced it would lend investors money against their newly purchased shares, in this case all 100 shares, not just the 20 paid for. Leverage on leverage is like using gasoline to light your barbecue. This raised the shares even higher and once it started going up, the mania of public buyers kept driving the stock to nosebleed heights, allowing the company to sell more shares at higher profits.

On New Year's Day in 1720, one share of the South Sea Company traded at 128 pounds, and by June 24, it had run to 1,050 pounds. Banks were lending people money backed by their shares of South Sea Company, which then bought more shares.

All bubbles end, and usually for no good reason except an economically unsustainable business is, well, unsustainable. As soon as the stock starts dropping, if only because it has exhausted its buyers, the whole thing unwinds. People who had borrowed money sold the stock to pay off the loans, which sent the stock price lower, the upside in exact reverse. Gravity works. In fact, Sir Isaac Newton himself dropped 20,000 pounds faster than an apple falls from a tree.

The South Sea Bubble is a great story, especially in relation to the Internet Bubble of 1999–2000, when shares of Priceline.com, which sold discount tickets on airlines such as Delta Airlines, were worth more than the airline companies themselves.

But it is the ramifications of the burst Bubble that is more telling. Banks which lent money went bankrupt. People hoarded gold and the British economy ground to a halt. Parliament passed the Bubble Act of 1720, which forbade speculation and required Acts of Parliament or the monarchy to create new joint-stock companies. Isn't it amazing how fast Parliament operated then?

While this may have solidified the government's grip on trade and kept mercantilism dominant, it had the opposite effect in the long run. Businessmen became so disgusted with government-mandated monopolies, that mercantilism probably peaked on that same June 24 day when South Sea hit 1,050 pounds. Adam Smith was born in 1723, and his book *Wealth of Nations* would not be published until 1776 at the dawn of the Industrial Revolution, but the memory of the South Sea bubble kept England in the muck for 50 years.

A minor footnote in the Bubble Act also forbade any company except for the Royal Exchange Assurance and London Assurance from writing marine insurance policies. Lloyd's of London slipped through a loophole since it wasn't a company, but a meeting place for individuals who underwrote insurance. I'll get to the implications of insurance in a coming chapter, but it added to England's aversion to risk.

In April of 1721, a plan was agreed on to fix the South Sea problem. Remember, the members of Parliament were complicit in the scheme; it was their debt that was being swapped for South Sea Company equity. The plan was called a consolidation. The government would swap shares of the South Sea Company for a piece of paper that would pay 3% interest forever. That's right, forever, a perpetual annuity that paid out interest year in and year out. In 1726 came the Three Per Cent Bank Annuity, and in 1751, the Consolidating Act gave Parliament the power to issue these instruments, or consols, short for consolidation. England was relieved of the burden of paying back big money in 1721, in exchange for an agreement to pay interest every year. They still pay interest on these consols, the interest rate fluctuating with the market.

The odd effect was that English investors became very risk averse. Burned by equity and now guaranteed interest payments forever, the store of wealth became the guarantee of interest, which put a dagger through the heart of risk capital. In the 1870s, the British were wealthy and America was poor. Undersea telegraphs opened America to British capital markets, and tons of money flowed to the U.S. to finance the building of the U.S. railroads. But rather than owning equity in these railroads, the British mainly lent money, still more comfortable with getting paid interest and their eventual money back, rather than take the risk in equity, 250 years after the South Sea Bubble burst. Even today, you find a much less developed venture capital business in England and all of Europe verses the risk junkies of Silicon Valley.

But the consols allowed England to raise capital, and never have to pay it back! The French never invented such an instrument after their Mississippi Company bubble, and were forced to borrow capital with very short duration, meaning they had to pay back their debts within a year. This long-term (perpetual) verses short-term debt capability would have serious consequences 70-plus years later during the Napoleonic Wars, favoring England in building its resources.

So while mercantilism and government intervention in trade peaked with the South Sea Company, risk capital was dealt a deadly blow as well. It would take the Industrial Revolution, happening away from government and away from the capital markets, to get things cranking again.

But the legacy of pre–Industrial England would set the stage for the Empire. The combination of the joint stock company, patents and its survival through a Bubble would eventually allow for the raising of risk capital again.

So how do you get paid for running an industrial engine and an empire? Very carefully. It's one thing to make cheap comfortable underwear and deliver it via railroad and steamship to a customer's doorstep, it's another thing to get properly paid for it, create lasting wealth, and maintain your lead.

By mid–19th century, England was running huge trade surpluses, exporting more in value than it was importing. Hey, that is what an industrial engine should be doing. The difference between exports and imports was made up in gold. Sort of a "if you want our goods, fork over the glitter." Gold (the Brits never trusted silver) was the only accepted form of payment and it became the only form of international settlement. England did anything to get it.

A glitch early in the 19th century should have tipped off England about the problems with gold. Protectionist Corn Laws cut off France and continental Europe from selling grain to England, which as gold drained out of their countries, was the only way to pay for British industrial goods. England never bothered to worry about the economic health of its customers.

If it were up to Parliament, the Industrial Revolution and the British Empire never would have happened. As only politicians can do, Parliament caved in to pressure to maintain the status quo and almost killed the free trade that would create the Empire.

During the Napoleonic Wars, imports from continental Europe, especially of foodstuffs such as corn and wheat, were limited. Farmers planted more of the crops to meet high prices. With victory, a flood of cheap corn and wheat began to flow into England. In 1804, Parliament passed a Corn Law putting duties on foreign corn, although this wasn't new; laws and duties of this sort had been imposed on and off since the late 17th century.

In 1815, with Napoleon truly defeated, wheat prices dropped by half. Landowners, who were the ones really represented in Parliament, fought for and won the passage of additional Corn Laws, which set a minimum price on wheat below which duties were imposed. By now, England had industrialized, and workers were jam-packed into cities with no ability to grow their own food. They protested the increase in the subsequent price of bread and demanded higher wages from factory owners. You can imagine factory owners scrambling to hire lobbyists of their own, but remember, they were predominantly Dissenters, with very limited representation.

A bad crop in 1816 sent prices flying high. Food riots became common as workers demanded higher wages to pay for food. In 1819, there was a massacre as troops opened fire on protesters in Peterloo in Manchester. Out of 80,000 protesters, 300–400 were killed as the Lancashire militia charged to tear down banners that read "No Corn Laws" and "Universal Suffrage." Pretty racy for 1819. It wouldn't be until the formation of the Anti-Corn-Law League in Manchester in 1839 that the movement against protectionism was formalized. The Corn Laws were finally struck down in 1846.

Sure, England prospered even with misguided protection-ism. But workers and therefore factory owners were under a lot of strain. No one in Parliament bothered to figure out the derivative consequences of protecting landowners, who had very little to do with the economic engine of the Empire. Worker discontent, of not only expensive food but lousy working and living conditions, would open the door to new political movements. Marxism and socialism would be born out of this discontent and the battles to remove shortsighted laws.

Economists like to point out that an ounce of gold represents the price of a man's suit, and has for thousands of years. That ounce of gold represents the cotton or wool, the creation of the thread and the cloth, the tailor cutting it to size and sewing it all together, and perhaps a modest profit. There's a lot going on there.

But while you can wear the suit, you can't do much with the gold, or four Ben Franklin greenbacks today. You've got to spend it on something to redeem its value. I know this is obvious, but it's still worth repeating. Money is just a con-duit. It is the value of work already done. Borrowing or credit is the opposite, the value of work to be done.

Spain discovered gold in the New World. OK, to be fair, the Aztecs had discovered gold and in 1519, Hernando Cortez and the Spanish discovered the Aztecs. This increased the world's known supply of gold, and like magic, economic activity and trade increased. Why? I suspect the Aztec gold was a shot in the arm, a bump in the world's money supply. Of course, there was no official trade bureau keeping track of such things, so the correlation was proba-bly not noticed. Spain faded into oblivion because all it did was steal gold, and rarely created enterprises. Other countries from France to England used the gold as a unit of money for transactions and built enterprises and robust economies, i.e. true wealth verses the transient kind.

Sherman, set the Wayback Machine to 1694 England (said Peabody) and let's see if gold is all that it is cracked up to be.

OK, we've been here before. Mercantilism is all the rage. The British monarchy has the crown jewels, but not much gold and not much liquidity. So joint-stock companies are formed to explore/exploit India and points east as well as the New World to the west. The Bank of England is established as the banker to the government, which, put a better way, means it would handle the government's debts.

As such, the Bank of England was in charge of issuing currency in the form of banknotes. It would trade a piece of paper that said 100 pounds for the equivalent value (not weight) of gold. The currency was backed by the gold in its reserves. This so-called commodity money sure beat carrying around heavy bars of gold for big-ticket items.

But hardly anyone ever came into the Bank to ask for the gold. It just sat around tarnishing. Goldsmiths in London had long before figured out they could lend money against the gold they held, making money on the interest payments. As long as they were careful to only write good loans, and not everyone asked for their gold back at the same time, this so-called "fractional reserve banking" was a hugely profitable business.

Well, the Bank of England existed to make money for its secretive shareholders. So it implemented this goldsmith sleight of hand, and jumped into fractional reserve banking in a big way, by issuing loans, sometimes for 10 times as much gold as it had on hand, mainly to the government. These loans were secured by future tax payments. The more gold in their reserves, then the more loans they could write. Mercantilism was the government's plan to stock those reserves with lots of gold, so it selfishly could borrow more.

The banknotes the Bank of England issued were known as fiduciary money, money based on trust. There were gold reserves backing the banknotes, sort of, and not enough if people really wanted the gold. More gold meant the money supply could increase as more of this fiduciary money was created. It wasn't a bad system, money supply had to come from somewhere. Using gold as a store of wealth was already a sleight of hand, money only fractionally backed by gold was a sleight of a sleight. So what. But then someone messed it all up.

Sir Isaac Newton had leveraged his planetary dreaming into a real job, as Master of the Mint. In 1717, the Newt decided that the Guinea would represent 129.4 grains of gold. Put another way, an ounce of gold was worth 4 pounds, 4 shillings and 11½ pence. You can tell Sir Isaac was a scientist, if he was an engineer he would have rounded it down to 4 pounds per ounce and called it a day.

He also set the price of silver at the same time and over-valued it relative to gold. Newton noticed that because of trade deficits, the East India Company was shipping most of England's silver to India to pay for tea and spices, and probably overvalued it on purpose. It didn't matter since holders of silver quickly sold it and bought gold for the onetime arbitrage, but it was the end for silver in England. Newton couldn't have known that silver had real industrial uses, that when formed into the compound silver halide it would be sensitive to light and usher in a huge photography business in Rochester, New York. Nor would he have known that the photographs would each be worth 1,000 words.

Newton's fixing of the exchange rate between money and gold was the market price in 1717. Fair enough. But like Newton's Third Law of Motion concerning momentum, that same exchange rate would last until 1931. It would take the world that long to embrace an alternative to the limiting gold standard. Until then, gold ruled.

David Hume argued in 1752 that more gold and more money supply didn't necessarily mean an increase in wealth for England. Ask Spain. He had a good point, but no one listened.

The Bank of England and other local banks issued fiduciary money, and in normal times, money supply was supposed to increase to the level needed in the economy. But if a profit was to be had loaning out new money by leveraging the gold, then the money supply would have to increase beyond what was needed. Econ 101 says too much money chasing too few goods means prices go up, i.e. that nasty word inflation. And this is in normal times. In abnormal times, all hell breaks loose.

That's what happened during the Napoleonic Wars. In 1797, there were rumors, which were not true, that French troops had invaded England. This caused a huge run on banks and financial panic, as everyone demanded their gold, to which the banknotes strongly suggested they were entitled. Knowing these gold reserves were needed to pay for the war, William Pitt as Prime Minister had the Bank of England suspend convertibility of its banknotes into gold, via the so-called Bank Restriction Act. Instead, the Bank issued banknotes backed only by the fiat, or command, of the government (maybe that's why those Italian Fiat cars rarely start up in the morning, no matter what commands you swear at them!) That's it. "We said thee is worth a pound sterling, so thee is, got a problem with that?" Fiat money was an interesting concept, the government could issue as much or as little money as it wanted to, gold be damned.

Between 1797 and 1821, during the Restriction, England's economy took off, partially because it was wartime, but also because the industrial age had begun. Steam engines ran mills. The Triangle Trade ran circles around everyone else in the world. Affordable English goods were in high demand as substitutes for homespun and homemade.

Manufacturers got wealthy, but landowners, who still controlled Parliament, were being left behind. One reason was that while English exports were growing, countries on the Continent, including postwar France, had nothing to pay for these goods with, except corn and grain. As we saw, Parliament, in the pocket of landowners, passed stricter and stricter Corn Laws. This had the effect of raising food prices for workers in factories, who demanded higher wages, but it also decreased the market for goods from these factories, because there was no way to pay for them, except with the grain that was more or less banned from England. How stupid is that? This was a double knock on both mercantilism and the gold standard.

This set up a huge debate between Bullionists, who demanded convertibility to check inflation, and anti-Bullionists, who argued against it. The anti-Bullionists conjectured that banks would only issue banknotes as merchants turned in their "bills of exchange," sort of like selling their accounts receivables. This was known as the Real Bills Doctrine, stating money was credit, and money supply would only grow to the level of actual credit in the system. John Law first developed it in 1705.

This was not good news for anti-Bullionists, even though they were probably right. Law was soon exiled after winning a duel and lived in Paris. While there, he became buddies with the Duke of Orleans. After the death of Louis XIV, the French monetary system was *une* mess, and with the Duke's help, Law volunteered to fix it. He set up Banque Generale, which issued fiat currency, backed by zip. And despite his Real Bills Doctrine, Law issued currency like it was, well, paper. Around the same time, he set up the Mississippi Company, whose stock ballooned concurrent with the English South Sea Company, eventually being worth more than all the gold and silver in France, which from the looks of the reserves of Banque Generale, was not very much. A bursting of the Mississippi Bubble in 1720 caused a run on the Banque and a depression in France for years to come. Almost three hundred years later, the French don't call their banks, Banques, but Credits, as in Credit Lyonnais.

John Law proved economists shouldn't be businessmen and his reputation killed the Real Bills Doctrine. Even when "invisible hand" Adam Smith backed Real Bills the Bullionists weren't swayed. Too bad. Real Bills was only slightly flawed in that it didn't check the amount of speculative loans a bank could issue, since loans are the source of bank profits. A floating reserve requirement, putting limits on fractional reserve banking in good times, could have fixed that flaw. Perhaps a Real Bills Doctrine could automate the creation of money supply today, in a modern non-gold standard world. But Reserve Bank chairmen have too much fun adjusting interest rates and turning on and off money supply at their whim to entertain the thought.

Inflation raged throughout the Restriction period, up until 1814, helping the Bullionists' argument that a strong gold standard would hold off inflation. But there was a war on, so the economy was working overtime to supply both the military and the regular economy, and it was hard to keep prices from going up.

• • •

The protectionist Corn Laws were inflationary. Workers demanded higher wages to pay for higher food costs. Add to that the unrestricted loans from banks, which increased the money supply, and it was no wonder that inflation was rampant during Napoleonic War England. Peel, who had implemented the Bank Restriction halting convertibility, knew something had to be done. But he couldn't pass anything through Parliament, to either cancel the Corn Laws, which would hurt landowners, or restrict bank loans, which would hurt bankers.

Peel turned to gold for the magic he needed to kill inflation. He put together a Bullion Committee filled with, you guessed it, Bullionists who pushed through a repeal of Restriction, meaning a return to convertibility. One respected member was William Huskisson, who had held many positions in government, including MP from Liverpool, where docks were brimming with industrial goods. There, he saw firsthand the benefits of free trade and lower duties, although it took him until 1828 to help repeal the Corn Laws. Huskisson also pushed hard for the building of railways to lower transportation costs and helped cut the red tape to create the Liverpool–Manchester line.

There is a strange twist to his story. On September 15, 1830, at the opening ceremonies for the line, he hitched a ride in the locomotive *Northumbrian,* which was a larger, improved version of George Stephenson's *Rocket.* When it stopped, he got off and crossed the tracks to chat with the Duke of Wellington, a national hero for his final defeat of Napoleon. Despite shouts to Huskisson to get out of the way, the *Rocket* ran him over, mangling his leg. George Stephenson himself loaded Huskisson into the *Rocket* and raced to get him medical help. When he died later in the day, Huskisson became one of the first railroad accident fatalities, literally run over by the industrialization he fought so hard for.

After Wellington whacked Napoleon and ended the war, like magic, a period of deflation, or dropping prices, occurred. This strengthened the anti-Bullionists' case: A fixed price of gold is no way to run an economic system; especially when the output drops in price. Nonetheless, a Resumption Act (they should have called it anti-Restriction) passed in 1819, and mandated convertibility by 1821. Memory of John Law's buffoonery cast a long shadow and convertibility was the damper on runaway money supply.

The Bank of England and other banks went back to fiduciary money, loaning out banknotes with fractional reserves of gold. Still, there was another nine years of deflation until 1830, most likely because the gold exchange rate had been fixed for the last century thanks to Sir Isaac and prices during the war had gotten out of whack and needed 15 full years to adjust.

How strange. In times of peace, banknotes are not trustworthy enough and must be backed by gold in reserves, but in times of war, when nobody trusts anyone, banknotes are based on pure fiat, trust out of thin air. Welcome to the backwards world of banking.

As an aside, it is bizarre that fractional reserve banks, which describe just about every bank in existence today, are basically bankrupt. A deposit in a bank is an asset for you and me, but it's a liability for the bank, since it owes us the money. But then the bank lends out the depositors' money as loans, and those loans are a bank's assets. Banks almost always have fewer "assets" than "liabilities," and therefore a negative net worth. And great profits until everyone wants his or her money back. So Trust with a capital T is key. It's the same for stock markets. If everyone sells Intel on the same day, its price would go to zero; there are not enough buyers or capital to handle a run on the stock. But it doesn't go to zero. At a low enough price, some investors see the value of Intel's future earnings and start buying it. So stock markets are built on Trust with a capital T as well, but there is a price mechanism to hold off runs and panics. Not so with banks.

More trade meant even more gold flowed into England, but unfortunately, there was no outlet for it. The Brits could have bought more foreign goods, which would have reduced their trade surplus and incoming gold, but there was not much to buy, and the echo of mercantilism of times past still encouraged exports and hoarding gold. England might have bought foreign fixed assets, maybe land or buildings, but they weren't necessarily for sale. Even if these assets were for sale, the legality of a foreign ownership or even getting legal title was questionable. The best solution to England's gold buildup would have been to use it to set up factories in France or Germany or America. Politically, this was a dead issue. No way was that going to happen.

So England was destined to suck up every last nugget of gold. Two major events kept this economic system alive well past its usefulness. Both happened around 1850. The first was the new discovery of gold in California, Australia and South Africa. The money supply to run the world's economy got the bump it needed to buy British goods. The other was the creation of screw propeller steamships, which lowered transportation costs by 50–70%, and increased demand for British products.

The more gold England had, the higher its bank reserves. With no outlet, this led to more banknotes in circulation, whether the economy needed that increased money supply or not (it usually didn't need it, there were enough Spinning Frames and Ironworks). Too much money chasing too few goods meant price inflation. Wages went up. Interest rates went up, another by-product of too much money and inflation, which often caused banks to fail, and resulted in runs on those banks. The fractional reserve banking system was anything but stable, all because too much gold was in the British banking system. England had become Spain, laden with gold and not enough to spend it on.

So England devised a way to get rid of gold. It turned out, at least in my opinion, to be the wrong way.

Bank runs and financial crises from too much gold became common: They occurred in 1825, 1847, 1857 and 1866. Think about it. In periods of inflation, money loses its value relative to the goods it is buying. This lack of faith in money causes people to move into real assets, including gold. Even though money was exchanged into gold at a fixed rate, the fear that the rate would change when the money lost value caused depositors to ask for real gold from banks. Plus, if a local bank failed, their banknotes would be worthless. Better to convert to gold quickly. Lack of faith is disastrous.

Something had to be done. For answers, most economists looked back to something David Hume had written back in 1752:

There seems to be a happy concurrence of causes in human affairs, which checks the growth of trade and riches, and hinders them from being confined entirely to one people; as might naturally at first be dreaded from the advantages of an established commerce. Where one nation has gotten the start of another in trade, it is very difficult for the latter to regain the ground it has lost; because of the superior industry and skill of the former, and the greater stocks, of which its merchants are possessed, and which enable them to trade on so much smaller profits. But these advantages are compensated, in some measure, by the low price of labour in every nation which has not an extensive commerce, and does not much abound in gold and silver. Manufactures, therefore gradually shift their places, leaving those countries and provinces which they have already enriched, and flying to others, whither they are allured by the cheapness of provisions and labour; till they have enriched these also, and are again banished by the same causes. And, in general, we may observe, that the dearness of every thing, from plenty of money, is a disadvantage, which attends an established commerce, and sets bounds to it in every country, by enabling the poorer states to undersell the richer in all foreign markets.

The best and brightest economists of the time met in Paris in 1867, to discuss a way to have both sound money and increased international trade. They came up with a system known as the "Price specie flow." Sounds like a case of the runs, rather than a cure for bank runs.

In 1870, even though England's economic power had already peaked (but who knew?), the bankers and government officials agreed to this system—better known as the classical gold standard—since the economists were promising them a system that would naturally balance trade and keep governments from screwing up by issuing too much or too little money. There were four "rules of the game."

1. Gold exchange rates for each country's currency is fixed
2. Gold is to move freely between countries
3. Money supply in each country is tied to the movement of gold, it goes up when gold moves in and down when gold moves out
4. Labor wages in each country are flexible

Hume continued:

Suppose four-fifths of all the money in GREAT BRITAIN to be annihilated in one night, and the nation reduced to the same condition, with regard to specie, as in the reigns of the HARRYS and EDWARDS, what would be the consequence? Must not the price of all labour and commodities sink in proportion, and every thing be sold as cheap as they were in those ages? What nation could then dispute with us in any foreign market, or pretend to navigate or to sell manufactures at the same price, which to us would afford sufficient profit? In how little time, therefore, must this bring back the money which we had lost, and raise us to the level of all the neighbouring nations? Where, after we have arrived, we immediately lose the advantage of the cheapness of labour and commodities; and the farther flowing in of money is stopped by our fulness and repletion.

Again, suppose, that all the money of GREAT BRITAIN were multiplied fivefold in a night, must not the contrary effect follow? Must not all labour and commodities rise to such an exorbitant height, that no neighbouring nations could afford to buy from us; while their commodities, on the other hand, became comparatively so cheap, that, in spite of all the laws which could be formed, they would be run in upon us, and our money flow out; till we fall to a level with foreigners, and lose

that great superiority of riches, which had laid us under such
disadvantages?

I think you get the concept. Each country would still use
its own commodity money backed by gold, but at that fixed
rate of exchange. The dollar was $1/20$ of an ounce of gold.
The British pound sterling was, thanks to Newton's precision, around but not quite $1/4$ of an ounce.

The classical gold standard was basically a check on
inflation, a way to make sure governments don't just print
money willy-nilly. No wonder economists invented it—it
gave them power.

The more gold and therefore the more money supply a
country had, the lower interest rates would be, initially anyway—too much money chasing too few borrowers. This
would stimulate domestic demand, entrepreneurs would
borrow money and build factories, wages would go up, and
consumers would borrow money against their future wages
and spend it. Inflation coming?

So if a country had lots of gold and overstimulated the economy and its consumers had lots of money to spend, it would
then import more than it exported. This country would then
have to ship gold out to make up the difference. Inflation and
gold leaving the country would drive up interest rates and slow
down borrowing and spending, until gold stopped flowing out.
Companies would then lower wages, making their products
more competitive, so they could export more and have gold flow
back into the country and then repeat the process all over again.

It's one of those brilliant concepts that works perfectly in
a vacuum, but breaks down in real life. The problem with
the classical gold standard is that the whole concept was
based on the competitiveness of workers' wages. When
some rich *flaneurs* (idlers, slackers) in France began buying

too many British pots and suits, and the gold flowed out of France, all the lower-class French workers had to take a pay cut or get laid off. That wasn't what they'd signed up for. No one puts up with a cut in wages without a fight. "But gold is flowing out" is a little tough to explain to the common worker. So they revolted and formed unions. Marxism, socialism and antiproductivity political regimes would soon flourish. All for some shiny metal.

The wrong thing was held constant. If wages had been held relatively constant and the exchange rate of money into gold allowed to float (like today), then workers would not have been as disaffected. In an uncompetitive country, instead of wages going down, the value of the currency would drop, import prices rise, and the blame laid on the foreigners for increasing prices. The flip side would have worked well for England. With a rising currency from a floating exchange rate, products like textiles would have gotten cheaper in both England and foreign markets, but not quite as cheap. So what? The market would still grow. National wealth would be created from a rising currency and money supply could grow at its natural Real Bills rate, rather than be affected by too much gold.

Rather than lowering wages and disenfranchising their customers, the British should have been working on ways to increase the wealth of all these other countries, because they were the end markets for the goods. And the more gold they collected, the smaller the markets became for their products. Pretty stupid.

The classical gold standard makes sense for a static world, one in which England trades lumber to France for wheat. It completely breaks down when industrial output is elastic. When the cost of iron goods and pottery and cloth and clothing is on a downward curve, making it more affordable for the masses and increasing the standard of living for everyone, it makes no sense to cut back production because you are TOO successful and gold flows into England. If England can offer a product cheaper than someone else can make it, then England ought to be able to sell as many as it possibly can.

The buyer will figure out a way to pay for it, almost by definition. If, because of volume production, England can sell a shirt for $5 that costs $15 for others to make, then buyers will line up around the block because they will save $10 for each and every shirt they buy. In fact, if you sold it for $14, you would do OK, but if you sold it for $5, you might sell a lot more shirts. That's what economists call elasticity.

They might even buy three shirts for every one that they used to buy. Either way, the buyer substitutes your cheaper shirt for the more expensive one, and has money to spend on something else. The old shirtmaker will have to find something else to do, but so what, his shirts are uncompetitive. But not because wages are too high, but because of the huge gap between hand made and factory produced.

The classical gold standard fixed the wrong problem. Success from selling $15 shirts for $5 meant gold moved into England, increasing money supply, and causing inflation. The inevitable increased wages theorized by the classical gold standard would make England's goods less competitive, until gold flowed out and trade was balanced. But that would in no way close the gap between the $15 handmade shirt and the $5 power-loomed shirt. But why penalize progress? In the end, France and Germany industrialized and killed off their own cottage industry anyway. And closed the gap. Price gaps are to be exploited, not closed, but all the classical gold standard could do was level the field.

The reason to spend so much thought on the gold standard and its holding back of money supply is that elasticity renders a gold standard useless. In fact, a gold standard becomes dangerous as it distorts the real market for products and holds back increases in living standards in two ways: (1) It hurts shirt producers by decreasing the available market for the $5 shirt seller, and (2) It hurts buyers as they must pay more to the old $15 shirtmaker, rather than buy the cheaper shirts and spend money on something else.

To be fair, England did have a problem. Bankers did lend out too much money in good times. It was profitable to do so and impossible to stop them. Regulation might have helped, but I doubt it. The Bank of England in 1870 and beyond needed a way to track money supply in the country, and make sure it didn't expand beyond a Real Bill–like pace. There were no computers to track such things, instead just those Dickensian men in visors and green eyeshades at the banks. It was trial and error. The information just wasn't available to determine an overheated inflationary economy until it was too late, so the classical gold standard was the less than ideal but workable solution.

Between 1865 and 1912, it is estimated, the capital markets in London raised 6 billion pounds. Of that total, 4 billion was invested outside of England, in "bond-based infrastructure ventures, such as Railways." Hmmm. By 1865, England no longer had any decent investments at home? Perhaps not—the Industrial Revolution had played out. Capital, backed of course by that accumulated gold, flowed out to lend money to American railroads. The echo from the South Sea still kept the Brits risk averse.

PART 3:

COMPONENTS NEEDED FOR COMPUTING

OK, enough about money, let's get back to computing. We are still searching in vain for components for Logic and Memory, not to mention dials and wheels and gears. Magnets will have to do for now.

During the sooty industrial era, the electricity of Volta and Galvani wasn't standing still. In 1819, Hans Christian Oersted, a Danish scientist, in a reflective moment postulated that there is a "unity of nature's forces," and that electricity and magnetism somehow must be related. He ran electricity through a loop of wire and noticed it was magnetized. How cool to be right!

In 1820, an Englishman named Michael Faraday made a living as a bookbinder and taught himself how to read by perusing his customers' books while binding them. He also dabbled in the sciences and had an inspiration similar to Oersted's, only he harnessed that magnetic force from electricity. By holding the loop of wire steady he got a magnet to move through it. Faraday had the first electric motor.

By 1826, another Englishman, William Sturgeon, wound wires around an iron bar and created an even stronger magnet, so strong that an American named John Henry would use these electromagnets to lift 1,000 pounds.

In 1830, John Henry, then a professor at the Albany

Academy, strung a mile of wire inside his classroom and ran an electric current down the wire to a small electromagnet that moved and rang a bell. You rang? In effect, he demonstrated the telegraph. He also invented the electromechanical relay, which would open or close a switch instead of ringing a bell.

By the late 1830s, Englishmen Charles Wheatstone and William Cooke and Americans Samuel Morse and Alfred Vail almost simultaneously commercialized the telegraph that Henry had demonstrated. They used simple electromagnets to create telegraphs for long-distance communication. A battery connected through a switch or key created current along telegraph wires, and moved electromagnets that got a marker to write on a strip of paper depending on whether the key was opened or closed. After a while, there was just too much paper flying around, so in the United States, the convention was to have a sender transmit a message using the key, and a receiver listen by ear to the clicks of a relay and write down the message. Vail invented a code with dots and dashes that could efficiently transmit words down the telegraph wire. Morse was the boss, hence it became known as Morse code.

Morse had to wait for congressional funding of $30,000 in 1842 to complete work on the telegraph. Morse used the same John Henry relays hooked to batteries as regenerators along the line, to get his telegraph signal to go more than 20 miles. This way, in 1844, he was able to run a 41-mile line between the U.S. Supreme Court in Washington, D.C., and the Mount Clair train station in Baltimore and on May 24 transmitted "What hath God wrought?"

Ezra Cornell was hired to run the original 41-mile line underground, but Morse couldn't get it to work; it probably shorted out from moisture. Cornell then ran wires on poles looped around glass doorknobs for insulation.

Cornell would wire up the East Coast with telegraph lines, go broke, buy out the bankrupt company, and merge it with competitor Hiram Sibley's company to form Western Union in 1851. With the completion of the St. Joseph, Missouri, to Sacramento telegraph line in 1861, the Pony Express was put to rest. Now Western Union operated coast to coast. By then, Western Union was in the control of John Pierpont Morgan and the Vanderbilts, who provided the expansion capital. Cornell took his wealth and "would found an institution" overlooking Cayuga Lake in upstate New York, which I was lucky enough to attend, and where I learned about Faraday and others.

By 1866, undersea telegraph cable would be laid off the back of the steamship *Great Eastern,* from the family of steamships built by I. K. Brunel (remember him) that made the initial transatlantic voyages. It increased the speed and lowered the cost of communications between the U.S. and Europe and would connect inventions and commerce between the two once separate worlds.

The unseen bonus of the telegraph was a cool little device known as the relay, basically a spring-loaded switch that an electromagnet could close if current ran through it or that the spring would open if there was no current in the electromagnet. It was pretty easy to build, a magnet, a few wire coils, a spring, no big deal. A relay was an interesting device because it had memory—it was either opened or closed. It turned out that the relay was the perfect logic element for computers, but not for another 100 years. Babbage should have been playing with these for his Difference Engine rather than gears.

Even 30 years after its invention, telegraphs could only transmit one message at a time, by opening and closing a circuit. A serious reward would go to the person who could get more than one message on a telegraph line. Alexander Graham Bell thought he had the answer. He was sitting in his living room when he noticed that when he struck a chord on a piano on one side of the room, the same chord would vibrate the strings on a piano on the other side of the room.

Bell conceptualized the harmonic telegraph that would send messages on different pitches or notes down the telegraph wire. He worked with the deaf (his mother was deaf) at Boston College and at night worked on the harmonic telegraph in 1873 and 1874. He found financing through Gardiner Hubbard, whose deaf daughter, Mabel, Bell tutored and dated, and George Sanders, whose five-year-old deaf son Bell taught to speak.

In March of 1875, Bell met with John Henry, then Secretary of the Smithsonian Institution down in Washington, D.C. Yep, the same telegraph-demonstrating-and-relay-inventing John Henry. Bell described not only the harmonic telegraph but also his dream of transmitting speech instead of dots and dashes. Henry told him to immediately drop work on the harmonic telegraph and focus on the speech transmitter, a "germ of a great invention."

But when he went back home, Hubbard told him to forget the telephone because there was already a lucrative market for the harmonic telegraph. And since he had invested money with Bell, Hubbard told him to back off Mabel. Sanders agreed with Hubbard—he was already in for $100,000 and looking for results. They also made him hire an assistant, Thomas Watson.

In June of 1875, Bell heard some sounds of piano strings through the telegraph. But this was impossible, since its relay simply opened and closed and couldn't transmit sound. But after taking the telegraph apart, he and Watson discovered that a contact screw was turned in too tightly in the relay, keeping it closed and keeping the current from the other side flowing continuously. This was a big break-through. But they still couldn't get it to transmit speech.

And then they got nervous. Bell heard about a guy named Elisha Gray, who had sold a chunk of his equipment company to Western Union Telegraph. Gray's company was named Western Electric Manufacturing Company and was busily working on a telephone. So Bell scrambled to file a patent for the telephone—the patent office had waved the requirement to have a working model. Bell filed his on February 14, 1876, a few hours before Gray. Then not three and a half weeks later, on March 10, 1876, Bell got his tele-phone to work with the famous: "Mr. Watson— Come here— I want to see you." Unfortunately, Bell had a really crude transmitter. It consisted of a metal rod moving up and down in a cup of acid whose resistance changed and thus captured sound and transmitted it down a line. Imagine sticking a cup of acid to your ear to talk to Grandma.

The Bell telephone worked, but was not yet practical. They worked on improvements, but not much helped, and within six months, Bell's backers (Hubbard was soon to be his father-in-law) tried to sell the patents to Western Union for $100,000. They said no. And then things took off.

In April of 1877, Thomas Alva Edison, an old telegraph keyer, patented a better transmitter, and best of all, with no acid. But so did a guy named Emile Berliner who beat Edi-son to the patent office by two weeks.

In July of 1877, Bell and his two backers formed the Bell Telephone Company and began using these improved trans-mitters. By September of 1877, a year and change after they declined to buy Bell's patents, Western Union jumped into the telephone business, cutting deals with Gray, Edison and a few others for equipment and patents. The year 1878 saw the introduction of the switchboard, ringer, and phone directory, really just a one page listing.

In November of 1879, Bell Telephone won a patent suit against Western Union, which gave Bell its entire phone business in exchange for 20% of phone rental receipts until the Bell patents expired. By early 1880, Bell had 133,000 telephones and lines and by the end of the next year, Bell's CEO Theodore Vail had bought enough stock of Western Electric to take it over.

In 1884, the first long-distance line went in, connecting Boston and Providence, Rhode Island. All of a sudden, Bell Telephone was in desperate need of something to amplify phone calls over these long distances. They had to go back in time to find the solution.

Power Generation

Before we fix A. G. Bell's problems, let's rewind to bookbinder Michael Faraday. He gave us electromagnets, which would go into early computers, but he also figured out how to generate power, which brought us light, which brought us . . . well, I don't want to get too far ahead.

In 1831, Faraday had the same unity thoughts as the Dane Oersted and imagined motors, but he also turned things around. If electricity could make a magnet move, why couldn't moving magnets make electricity? This simple thought of electromagnetic induction was the genesis of electric generators. In 35 years, generators would be better than Voltaic Batteries at creating electricity.

But while Faraday could move a coil of wire through magnets, the magnets were not strong enough to generate much electricity. In 1866, Charles Wheatstone would add another invention to his belt. He and German-born Brit Wilhelm Siemens would invent a generator that used electromagnets. It was basically a coil of wire moved through electromagnets. But because there was some residual magnetism after getting it started, some of the electricity generated would be used to make the electromagnets stronger. This amplified the magnetism, so it could generate more electricity. It was one of those rare devices that worked better the more it worked. It took external power and LOTS of efficiently generated electricity. This self-excited dynamo (sounds like someone who can't sit still in class) opened the world to large-scale power.

Here is where steam engines came back into play. The same steam engines that ran factories and locomotives and steamships could now drive electric generators and create huge amounts of electricity.

Edison used steam engines to generate electricity in New York City. Over time, flowing water, which oddly steam engines had displaced for power in factories, was best used to spin turbines.

When high-voltage electricity was connected to two pieces of carbon held close together, it gave off a brilliant light, an arc in the gap between them. By 1875, London, New York and Paris streets were being lit at night with these arc lamps, driven by dynamos, which quickly replaced dangerous gas lamps. So bright was the set of arc lamps that lit up Broadway, it was nicknamed the Great White Way.

Unless you were good at putting out fires, you probably didn't want to use an arc lamp inside your home. Arc lamps produced too much light, although they are still used today in spotlights for concerts and Broadway shows. But a modest light would be nice. In 1878, English scientist Joseph Swan found he could get thin wires of carbonized cotton, or filaments, to not arc, but to glow. But they would quickly burn out, as the filaments were exposed to oxygen in the air, and would get hot enough to spontaneously combust. So he put the filament inside a glass tube and created a vacuum, basically by sucking the air out of the tube. Now when he hooked it up to a battery the filament would glow and not burn out. Swan didn't consider it much of an invention and never patented the idea.

In October of 1879, Charles Batchelor, who was one of Thomas Alva Edison's assistants, demonstrated the same principle at his lab in Menlo Park, New Jersey. Once again, we had almost simultaneous inventions. Turns out that in 1876, Edison had bought out a patent by a Canadian named Woodward, but no matter, he had his light bulb and took out his own patent. John Pierpont Morgan and the Vanderbilts put up $300,000 in exchange for patent rights and capitalized the Edison Electric Light Company. Share prices of gas light companies would go up and down based on Edison's press releases. Edison and Swan fought in British courts for the rights to light England, and eventually created the Edison and Swan United Electric Light Company. Because of this joint venture, Edison tried to figure out another way to profit from his invention.

Edison knew plenty about electromagnets from his years as a telegraph keyer. He would often take apart telegraph tickers and put them back together in better shape. When he heard about dynamos, he had to build his own. In 1882, he installed a 27-ton direct current generator on Pearl Street in lower Manhattan to light the financial district, including J. P. Morgan's offices. His business model, which attracted the Morgan money, was to make money selling electricity, rather than selling light bulbs.

At the same time Edison was installing his DC generator in New York, a 29-year-old Serbian named Nikola Tesla had come up with plans for an efficient alternating current, or AC, generator. Unlike DC with its constant voltage, voltage swings from plus to minus with AC. With Direct Current, an electron has to move all the way down a wire while with Alternating Current, electrons just jostled around. With AC, energy comes from the constant changing of voltage from positive to negative, rather than moving electrons through wires. Because of this, AC would prove to be a much better way to distribute electricity over long distances—DC could go only about a mile. Tesla came to the U.S. in 1884 and pitched the idea of an AC generator to Edison. Too late, Edison had already installed DC—as stock traders might say, "he was seriously long DC"—although Edison needed to fix the design of his own generator. So he offered Tesla $50,000 to fix it, which Tesla did and then Edison stiffed him for the 50 large. But in the meantime, Tesla had filed for a patent for his polyphase AC system, one of over 700 patents he would receive in his lifetime, including a few on arc lamps.

Tesla ended up hooking up with George Westinghouse, who was intrigued by AC power. In May of 1885, Westinghouse bought the Tesla polyphase patent for $5,000 in cash and 150 shares of Westinghouse stock. But Tesla also got royalties of $2.50 for each horsepower of AC generating equipment installed, which was potentially worth billions. Edison constantly badmouthed AC generators and Tesla. Edison, a fierce competitor, claimed AC was unsafe, which was not true. He went so far as to call a press conference in 1887 where he showed a Westinghouse generator zapping neighborhood dogs and cats. Nice neighbor. It was called execution by electricity or electrocution. New York State adapted the method a year later for capital punishment.

J. P. Morgan and Westinghouse fought a mean battle for control of electricity distribution. Edison and Morgan continued to spread stories to investors that AC was unsafe and that Westinghouse would never make money because of the Tesla royalty payments, and that was true. To help Westinghouse get funding from the market, Tesla gave up his royalties.

Edison Electric, which would eventually turn into General Electric, charged $6 per streetlight installed, and Morgan pumped money in to build power. But so did Westinghouse, who showed off the Tesla AC system at the 1893 World's Columbian Exposition in Chicago, the World's Fair of its time. This got Westinghouse the contract to build two massive generators spun by Niagara Falls in 1895. Slowly they turned, inch-by-inch, step-by-step, until General Electric eventually switched to AC and made the next 11 generators for Niagara Falls. These generators would quickly supply power for Buffalo and many of upstate New York's cities and factories.

Those Niagara Falls generators were expensive. Entrepreneurs in shacks, even Thomas Edison, couldn't afford the huge capital costs to build a power grid in the United States. Wall Street was there for a reason, to provide this risk capital, to fund big growth projects.

Hey, wait a second. This book is about computers and integrated circuits, not lights and power systems. True, but it turns out that both Edison and Tesla made huge contributions to the computer industry, which wouldn't even exist for another 50-plus years.

Remember that Swan/Edison light bulb? It was nothing more than two wires attached to a thin filament that glowed inside a vacuum bulb. Over time, carbon deposits from filament would darken the glass bulb. In 1883, Edison worked on getting rid of the carbon by putting a metal plate inside the bulb. He put a positive charge on the plate, figuring it would attract the carbon. The carbon still sprayed around, but Edison noted that when he put a positive charge on the plate, a current would flow and if he put a negative charge on the plate, no current flowed. He named it the Edison effect (what else?) but promptly forgot about it. He shouldn't have.

PART 4:

DIGITAL COMPUTERS

Ballistics, Codes and Bombs

While working on alternating current, Tesla would wind coils and change the frequency of the current to see what would happen. To his delight, while out in the deserts of Colorado Springs in 1899, he got electricity to jump from these Tesla coils and light up the sky, sometimes in arcs of 100 or more feet. He also sent electricity wirelessly, lighting 200 lights 25 miles away. Of course, you wouldn't want to be standing in the path while he was doing this, but no matter. He had discovered that air could conduct electricity, and by increasing the frequency and turning down the power, these same Tesla coils could transmit radio signals. Tesla quickly filed patents.

He returned to New York in 1900. J. P. Morgan, who realized that Tesla was right and Edison wrong about AC/DC, fronted Tesla $150,000 in exchange for half the rights to Tesla's communications patents. Tesla then built a radio broadcast tower on Long Island, but never could get it to work. Morgan pulled Tesla's funding when he needed to bail Wall Street out of the Panic of 1907.

Meanwhile, on December 12, 1901, Italian Guglielmo Marconi, flying an antenna on a kite in Newfoundland, tuned in radio signals broadcast from 2,000 miles away in

Prolhu, England. British engineer John Ambrose Fleming, an old Edison employee, designed the transmitter. Edison claimed Marconi could never get his transmitter to work that far because of the curvature of the earth, not knowing (no one did) that radio signals would bounce off the ionosphere.

Marconi filed for patents, setting him up for a battle with Tesla, taking over for Edison! (The U.S. Supreme Court would eventually overturn the Marconi patent in favor of Tesla's in 1943, around the time of Tesla's death.)

The Tesla coil would go to great use in radios, TVs, radar, X-rays and fluorescent lights. He even fashioned a death ray that the FBI would eventually be interested in during World War II. But Tesla moved on and kept filing more patents.

The most curious were from 1903: Method of Signaling 723,188 and System of Signaling 725,605, sometimes referred to as the AND-logic gate circuit patents. These two patents described a basic logic element needed to implement Boolean logic, set up by our old friend, George Boole. Finally.

An AND-logic gate is quite simple. It has two inputs and one output. If both inputs are ON, then the output is ON (1 AND 1 = 1). But if either input is OFF or both are OFF, then the output is OFF (1 AND 0 = 0).

In fact, with 4 AND-logic gates, you could create a memory element, known as a flip-flop. So with AND-logic gates you could create both the Logic and the Memory elements needed to make a computing device. Pascal is turning over in his grave.

Today, every computer and microprocessor is made up of millions of these logic gates. Tesla nailed it back in 1903. A lot of good it did him, because there was no way to really implement AND-logic gates.

Oddly, it would be Tesla's old nemesis Edison who would accidentally trip across the way to implement AND gates, and more effective radio transmitters and receivers, without even knowing it: the Edison effect.

This was around the same time that Herman Hollerith was inventing punched cards to tabulate the 1890 census, remember that?

Hollerith figured out that businesses also do a lot of counting, or accounting, so he formed the Tabulating Machine Company in 1896 and discovered a huge market for his tabulators. Business boomed and by that time, the Tabulating Machine Company had morphed into International Business Machines. These were still mechanical machines; electricity was being used to power lights and radios but not yet computing.

John Ambrose Fleming, Marconi's transmitter designer, had previously worked for the Edison Electric Light Company in London. There he became familiar with the Edison effect and solved the mystery that Edison never pondered. Fleming postulated that it was electrons that flowed from the filament, attracted to the positive charged plate, but repelled by a negative charged plate. Current would only flow in one direction, like a valve. But Fleming went further. Edison hated alternating current, remember, and never bothered with it. Fleming, on the other hand, was intrigued with AC, and in 1883, applied alternating current to the filament and noted that direct current came out at the plate. A device of this sort is known as a rectifier or a diode, and is the Holy Grail for radio transmitters and receivers. But it would be 20 years before Fleming would realize why someone would want to use a light bulb in a radio. In Fleming's own words:

It was about 5 o'clock in the evening when the apparatus was completed. I was, of course, most anxious to test it without further loss of time. We set the two circuits some distance apart in the laboratory, and I started the oscillations in the primary circuit. To my delight I saw that the needle of the galvanometer indicated a steady direct current passing through, and found that we had in this peculiar kind of electric lamp a solution of the problem of rectifying high-frequency wireless currents. The missing link in wireless was "found" and it was an electric lamp! I saw at once that the metal plate should be replaced by a metal cylinder enclosing the whole filament, so as to collect all the electrons projected from it. I accordingly had many carbon filament lamps made with metal cylinders and used them for rectifying the high-frequency currents of wireless telegraphy. This instrument I named an oscillation valve. It was at once found to be of value in wireless telegraphy, the mirror galvanometer that I used being replaced by an ordinary telephone, a replacement that could be made with advantage in those days when the spark system of wireless telegraphy was employed. In this form my valve was somewhat extensively used by Marconi's Telegraph Company as a detector of wireless waves. I applied for a patent in Great Britain on November 16, 1904.

In the United States, where everyone is so literal, the Fleming valve became known as the vacuum tube, and it unleashed the radio industry by allowing cheap transmitters and receivers.

Two years later, American Lee De Forest modified the Fleming valve by adding a third element, called a grid, between the filament and the plate. The grid controlled the flow of the electron between the two other elements. What is so amazing is that the De Forest "triode" could act as an amplifier, or it could just be an electrically controlled on/off switch, like a relay but without the moving parts.

In 1906, its most important use was as a radio receiver. In 1907, De Forest patented the triode and called it the Audion. This triode vacuum tube could be used as an oscillator and operate as a radio transmitter. Radio mania took hold for the next 20 years. Everyone overlooked one subtle use for the triode as a switch, the AND-logic gate. It would be the perfect logic element. AT&T ended up buying rights to the De Forest patent so they could improve the design for amplifying long-distance signals.

So punched cards and tabulators were the state of the art in computing for the first 40 years of the 20th century. Electronics were revolutionizing communications but not computing. The Radio Corporation of America was booming, and vacuum tube radios were flying off the shelves to receive music, news, Amos and Andy. Then in 1926, Philo Farnsworth figured out that he could make a larger vacuum tube, and use electromagnets to control the electron beam to scan phosphorous painted on one end of the tube. He invented television.

But there was no compelling need for computers, no huge calculations that couldn't be done by hand, so there was no huge rush to turn vacuum tubes into computers. Into the 1900s, operators ran most of the telephone system. When you asked for Murray Hill 7-0700, an operator would run a patch cord between your line and 0700. American Telephone and Telegraph was busy building its nationwide telephone network, using vacuum tubes as amplifiers on long-distance lines. The company never bothered using vacuum tubes or even relays as switches. Operators were cheap.

At the turn of the century, Automatic Electric Company,

one of only a few competitors of Western Electric, was making telephone equipment, including dial phones that operated a mechanical switch called Step by Step. Almon Strowger, an undertaker, had designed and patented the automatic switch when he realized that his competitor's wife, the local phone operator, was siphoning away potential customers. It took AT&T until 1921 to begin using automatic dialing switches and replacing local operators. Then in 1938, it rolled out crossbar switches, big ugly devices that moved horizontal and vertical bars. At the same time, Bell Labs, the research arm of AT&T and Western Electric, was playing with relays and vacuum tubes to do switching. All of a sudden, electronic computing was on the not too distant horizon. The needs of the phone network would drive computing forward for years to come.

In 1937, George Stibitz, an AT&T mathematician, was doing calculations to improve the quality of long-distance calls and needed to calculate the physics of the magneto-mechanics of all the relays in the network. This involved some nasty computations, which he figured he could automate. He thought the same relays he was studying could help in the computation, and fashioned a binary adder, using a tobacco tin, of all things, as a switch to input numbers. He called it the Model K since he built it on his kitchen table. Stibitz received a patent for this device (the adder, not the table). In effect, he used the relays as AND-logic gates. The beauty of his approach was its simplicity; if you could build one binary adder, you could just hook up a few more and before you knew it, you could work with really large numbers, sort of a tobacco tin Pascaline.

By November of 1939, Stibitz and Bell Labs researcher S. B. Williams created the Complex Number Calculator, with 450 relays and 10 crossbar switches. The CNC, later renamed the Model 1 Relay Computer, could multiply two 8-digit complex numbers in 30 seconds. Complex numbers were critical in calculations for physics.

Did you hear the fireworks go off? The Model 1 was a logic machine, a glorified calculator, all logic, and no memory, all bat, no glove. But it is considered the first electronic digital computer, and that made Stibitz the father of computing.

Of course, there was a shadow effort going on at the same time. Seems like there always is. In 1936, John Vincent Atanasoff at Iowa State University was trying to solve differential equations. He could have used a bulky mechanical calculator, designed by Vannevar Bush in 1931, that was the long elusive implementation of the Babbage Differential Engine. But it didn't work very well; it was slow, bulky and broke down all the time. Atanasoff tried hooking up off-the-shelf mechanical products to automate calculations.

Then Atanasoff used a Monroe calculator, an accountant's machine that used the same stepped drum as the Leibniz multiplier did way back in 1694. He connected it to a Hollerith/IBM tabulator to sum results but he couldn't get them to work together. So he built his own from scratch. With his assistant Clifford Berry, he built the Atanasoff Berry Computer or ABC and demonstrated it in December 1939. It used vacuum tubes as switches, and condensers (a battery-like device that could hold a charge) for memory. The ABC was probably the first real electronic digital computer, but unfortunately for Atanasoff, World War II broke out, he lost his funding and the ABC gathered dust.

Here is where the story gets interesting and starts to resemble the birth of the steam engine and the Industrial Revolution.

The United States Army and Navy each had a problem. There was a war on, and they were painstakingly assembling artillery-firing tables, which were a series of complex, differential equations. The ballistic computations took in all sorts of inputs, length of weapon, size of ballistic, temperature, wind direction, wind speed, elevation and who knows what else. These tables were carried out in the field and the artillery crew would look up the proper and most accurate firing angle for their weapons to hit targets.

At that time, Howard Aiken conceptualized a program-controlled computer while getting his doctorate at Harvard. With $100,000 of funding from IBM, he began work in 1939 on the Harvard Mark I, eventually known as the IBM Automatic Sequence-Controlled Calculator. It had 3,300 relays and the rest was mechanical, but that didn't stop the Bureau of Ships from co-opting the machine to compute firing tables. The Mark I had 750,000 parts, weighed five tons and could do three operations per second. It constantly broke down. The Mark II, III and IV would eventually be built. Grace Hopper was given the task of programming these machines. When trying to figure out why one of her programs didn't work, she actually found a moth caught in one of the relays. No myth, she pasted the moth into the logbook of the Mark II (now at the Smithsonian). Hence the name computer bug.

Aiken would remark in 1947 that he thought six computers would just about do it for all the computing needs of the United States. The military should have set him straight. The Army, actually the Aberdeen Proving Ground Ballistics Research Laboratory in Maryland, was using a version of the Vannevar Bush Differential Analyzer to compute firing tables. Their "computers" were a hundred women in a room operating desktop calculators and feeding numbers in the Analyzer. It was not only slow, but also prone to human error, with numbers out of order or transposed digits. When Aberdeen ran out of capacity, the Army hired the nearby Moore School of Electrical Engineering at the University of Pennsylvania, mainly because the Moore School had its own Differential Analyzer.

A physics professor at the school, John Mauchly, had seen the Atanasoff ABC computer in June of 1940, and thought the Moore School could implement something along those lines. In August of 1942, Mauchly suggested to Lt. Herman Goldstine at Aberdeen that a vacuum tube machine would do calculations faster and more accurately. John Presper Eckert and Goldstine began building just such a machine, Project PX, the Electronic Numerical Integrator and Calculator, in mid-1943.

Even with fast computers figuring out firing tables, most munitions fired in wartime miss. Whether they are artillery or bombs dropped from planes, an estimated 90% or more are wasted; they simply miss their intended target. Of course, the solution in most wars is to either fire more munitions, or fire munitions that make bigger explosions to make up for the inaccuracy. Of course, accuracy is cheaper and more effective. Electronics would play a role in the precision of weapons beyond World War II and become the cornerstone of the American military doctrine in the 21st century.

Meanwhile, the Germans had similar computing needs but it seems less for artillery and more for the design of aircraft. A German engineer named Konrad Zuse worked diligently between 1935 and 1938 on a mechanical computer to solve simultaneous equations with 30 unknown variables—really hairy stuff. The Versuchsmodell 1 (V-1, or after the war, Z-1) just barely worked. It operated with a hand crank. He worked on another, the Z-2, which used mechanical memory, but relays for logic, although it didn't work that well either. By 1941, with a war on, the Nazi regime got interested in computers. A friend of Zuse and a member of the Nazi Party, Helmut Schreyer, got funding for a relay-only computer for the Aerodynamics Research Institute. In May of 1941, Zuse got a new Z-3 model working. It used 800 relays for logic and another 1,600 relays for memory and other control functions. Zuse even invented a programming language named Plankalkül to run the Z-3. A program would be read in on punched cards or tape, and executed by the Z-3. It was an early, if not the first, program-controlled computer. Perhaps the Z-3 did computations for aircraft design. Not much is known of the Z-3's use in the war effort, as the Z-3 was destroyed by Allied air raids in 1945.

Back on June 15, 1944, seventy-five brand-new B-29 "Super-fortress" bombers, manufactured by Boeing in Wichita, Kansas, took off from bases near Chengtu, China. Their target was the Imperial Iron and Steel Works in Yawata, Japan, a major supplier of armaments for Japanese battleships and tanks. It was the first bombing run aimed directly at Japan since the aptly named Doolittle bombing raid over Tokyo in 1942, and just nine days after D-Day in Normandy. Imperial Iron churned out some 2 million metric tons of steel each year, almost a quarter of Japanese wartime output. Taking out even one of the coke ovens at Imperial could shorten the war by months.

Remember, Englishman Abraham Darby invented the coke oven in 1710. And when the King of England needed cannons to put down the insurgency of colonists in America in 1774, cannon-maker John Wilkinson perfected these same coke ovens by adapting James Watt's steam engine to run his bellows. The need for cannons helped spark the

Industrial Revolution, and 175 years later, wars were STILL being fought over coke ovens.

The Superfortress had pressurized cabins, allowing it to fly at high altitudes and withstand the 15-hour round trip to Yawata. One B-29 of the 468th Bomb Group crashed on takeoff and was pushed to the side of the runway. Ten others aborted during takeoff. Sixty-four B-29s, each with a crew of 10, made their way toward Yawata. A few got lost, others dropped their bombs early and returned. A total of 47 bombers made it to Yawata, dropping 376 bombs, 500 pounds each. Despite this remarkable tonnage, the raid was one big dud. Only one bomb, ¼ of 1%, hit anything—a power station that wasn't even a target, three quarters of a mile away from the coke ovens. The other 375 bombs missed completely, taking out rice paddies for all we know.

The closest World War II came to precision bombs was a demonstration by behaviorist B. F. Skinner, that's Burrhus Frederic to only his closest friends. Working at the University of Minnesota, Skinner was oddly funded by food company General Mills. Bigwigs at the U.S. Office of Scientific Research and Development watched in amazement as the nose cone of a bomb accurately tracked a simulated Japanese destroyer. When they opened up the device, they saw three pigeons, trained to identify silhouettes of Japanese warships, happily pecking away on a window that provided feedback to guidance controls. (Is this how they get the holes in Cheerios?) Skinner trained the pigeons by rewarding them with seeds when they correctly pecked on the ship silhouettes. He found the pigeons responded better when he rewarded them with hemp seeds rather than grain seeds. Perhaps General Mills was contemplating a bid for Twinkies.

But the Pentagon was deep into the Manhattan Project and their secret A-bomb, and turned Skinner away.

• • •

The Allies didn't forget about the coke ovens. For the next 14 months, Yawata and the Imperial Iron and Steel Works remained targeted for destruction. The Allies tried bigger bombs, then incendiary bombs in an attempt to burn the city and the factory. Still, Imperial turned out steel.

On August 9, 1945, three days after the B-29 *Enola Gay* dropped the Fat Boy atomic bomb on Hiroshima, an arsenal located near Yawata was to be the target of the second atomic bomb; perhaps that would burn down the city and factory. But smoke from yet another B-29 incendiary bombing raid, trying to destroy the steel factory, obscured the arsenal. So the pilot of the B-29 *Bockscar* looked up his secondary target and headed toward the city of Nagasaki, killing 35,000 Japanese in the blink of an eye. The war ended a few days later.

It is Hiroshima and Nagasaki that are remembered. But that failed night raid over Yawata 14 months before marked the beginning of the end of the Industrial Age. Had the Allies possessed precision guidance systems it is quite possible that Hiroshima would never have happened. But lacking the ability to target ordnance with any precision, the only available solution was to make the bombs bigger and bigger until they became so destructive they were effectively impossible to use. But it soon became clear that atomic and nuclear weapons in themselves would produce only stalemate; real wars would continue to be fought by other means.

Accurately aimed bombs and munitions help win wars, but military intelligence tells you where to aim them in the first place. Important Nazi communications were encoded with a secret cipher. The Enigma code was developed at the end of World War I to encrypt financial transactions, but was never really used that way. The Enigma machine was a series of keys that sent electrical charges to rotating wheels. When it was first tested in 1932, the Poles cracked it and built one themselves. But by 1939, the Nazis got smart and changed the code every day instead of every month, and the Poles could no longer decipher the messages. But they did sneak their model of Enigma, which they named Bomba, to the British, which gave them a huge head start in breaking Nazi codes.

Alan Turing, a mathematician from Cambridge and then Princeton, had written a thesis in 1934 on a Universal Machine that could figure out any algorithm depending on how it was programmed. In other words, he conceptualized the stored-program computer with programmable instructions, rather than a fixed-purpose machine.

When war broke out, Turing, who had returned to Cambridge, was sent to the infamous Hut 8 at Bletchley Park to help break codes. By January 1940, with the help of Bomba, the Enigma code was broken. But the decoding work was all done by hand, which meant it would take several days to break the daily code, and by then, there was a huge pile of messages waiting to be deciphered.

Turing figured out a shortcut. By "disproving the validity of conjectures by contradiction," he reduced significantly the number of calculations and that allowed him to automate the deciphering process. In March 1940, Turing constructed the Bombe, a relay-based machine that helped speed up the daily cracking of Enigma codes. It wasn't so much a computer as a pure logic device specifically built to crack daily Enigma codes. By December 1944, there were 192 of these Bombes operating around Bletchley.

Hitler and his high command insisted on a tougher code, named Lorenz, although the Brits called it Tunny. Before D-Day, they had cracked the Tunny code, but instead of taking a day, which they had gotten the Enigma cracking job down to, it might take several weeks to crack each change of the Lorenz code. Irving John Good and Don Michie joined an existing group of Hut 8 scientists, Max Newman, Wynn-Williams and Turing and they constructed a device in April 1943 called Robinson, named after cartoonist Heath Robinson who drew strange machines; the British version of Rube Goldberg. The Robinson used relays and other electronics and required two sets of paper tapes to read 2,000 characters a second. Progress was slow as the tape kept ripping.

Two London-based post office engineers, Tommy Flowers and Alan Coombes, improved on the Robinson, more than doubling the speed, by using 2,400 vacuum tubes and a number of servomotors, instead of relays and fragile paper tape. But it was still a Turing machine. Its name was Colossus, a mighty name, and it was moved to Bletchley Park and started operating in December 1943. It had both logic and memory and since it could change its program based on which code needed to be deciphered, it was a true Turing programmable computer. The Allies relied on the Colossus to help determine Nazi troop concentrations and the best D-Day landing points.

But in 1945, after VE and VJ days and the end of the war, we all know what happened next. With the German Z-3 destroyed by Allied bombing and the Moore School's Electronic Numerical Integrator and Calculator, or ENIAC, still being assembled, the Brits owned the computer business. Turing showed the Colossus architecture to every university and industrial company in England. Each built its own version.

The British had a lead in computers, much as they had a lead in steam engines. Over time, all the parts needed for the Colossus class computers would be made in England and other parts of the sprawling Empire.

Sir William Schokleford won the Nobel Prize for an all-electronic tri-valve. Soon, various companies figured out how to make tri-valve integrated systems, or Trivits, which were oddly nicknamed "fries." The technology industry grew up just outside of Wimbledon. An Oxford dropout, Sir William Spencer Hanover Tudor Gatesford started EverSo-Bitsysoft and became the richest man in the world. The Digital Revolution happened in England, extending the Empire into perpetuity. The sun still doesn't set on that Empire, and tea is at 4:00. See you there.

OK, my bad. None of that happened. Instead, when the war ended, the British were scared stiff that Russian spies would infiltrate Bletchley Park and steal the design to the Colossus for their own secret intelligence. They destroyed all 10 Colossus machines and burned the plans. Some believe two of the machines were used by the British government for a few years, probably the tax department, and then destroyed. The existence of the Colossus was classified until the 1970s.

What? Destroyed? This turned out to be a costly mistake. Because of Turing, they had a shot at repeating the steam engine story. These computers were no different than steam engines, they added value to raw materials, in this case knowledge instead of iron or cotton. But some paranoid bureaucrat cost Britain its shot at leading the coming computer, semiconductor and communications revolution. It's as if they had destroyed John Wilkinson's cannon barrel boring tool after Napoleon's defeat. From here on in the saga, you won't hear too many British names or inventions. Amazing.

• • •

Two other important technologies, which used the Tesla coil and vacuum tubes as amplifiers, came out of World War II: RADAR for RAdio Detection And Ranging and SONAR for SOund NAvigation and Ranging.

Turing, who would later go to the University of Manchester, worked on the Manchester Automatic Digital Machine, or MADAM, and became famous for a posthumously published paper titled "Intelligent Machinery." In it, he outlined the Turing Test. A computer would most surely be intelligent if a human who fed it questions from the other side of the wall couldn't distinguish between it and a human answering the questions. Turing was convinced one could be built by the year 2000. Maybe. My bank's ATM is smarter than its tellers, and might actually pass the Turing Test.

Meanwhile, back in Philadelphia, things were moving kind of slow. Project PX, the Electronic Numerical Integrator and Calculator, or ENIAC, was started in mid-1943. Perhaps it was a little ambitious. It contained 17,468 vacuum tubes, 70,000 resistors, 10,000 capacitors, 6,000 manual switches, and 5 million solder joints. It weighed 30 tons and consumed 174,000 watts of power. Literally, lights would dim when this puppy fired up. The first program ran in February of 1946. The ENIAC was a little too late to win the war, but right on time to launch an economic revolution and win the next one.

At 5,000 operations per second, the ENIAC was fast. But of course it was no longer needed for ballistic tables. Good thing, because it failed on average every 2.5 hours. It also might take hours or even several days to change the algorithm or program that the ENIAC worked on. It had very little internal memory. Of course, the biggest problem with the ENIAC was that it was still a decimal machine working with 10 digits instead of the two of Boolean binary math. That increased its complexity, probably 100-fold.

One of the folks working on ENIAC was John von Neumann, who had come over in June 1944 from Princeton's Institute of Advanced Study, where Turing had studied. Von Neumann reengineered the ENIAC to store the algorithm/program inside it along with the data to be processed, and also added a "conditional control transfer." For memory, von Neumann noticed that mercury delay lines, used in radar systems to store aircraft location information, stored a pulse or wave in a vial of slow moving mercury. He quickly used these for his memory. And von Neumann insisted on a binary machine. The product of these ideas was the design of another machine, the Electronic Discrete Variable Automatic Computer, EDVAC.

The result, known as von Neumann computer architecture, consisted of a central processing unit that read in a program and data from memory and wrote the results back to memory. Virtually all computers in use today are von Neumann machines. Of course, von Neumann probably read Alan Turing's paper back in 1934, and knew about the Universal Machine. Since Turing had been studying at Princeton, however, he may have picked up on the concept from meetings with von Neumann. Today, both seem to share credit.

Since von Neumann worked as a consultant to the Manhattan Project, his computer expertise did play a part in the war effort despite the lateness of the ENIAC. He spent time at Los Alamos National Laboratory helping guys like Nick Metropolis and Richard Feynmann, who were using and often repairing electromechanical computers, to compute their physics calculations of the atomic bomb.

Just before the end of the war, the Navy funded a

computer-based flight simulator to train its pilots. The MIT Servomechanisms Lab received close to $1 million for Project Whirlwind (what a great name!). It was not terribly successful, but it was the first to use tiny electromagnets as memory instead of mercury delay lines.

Unlike the British who destroyed the Colossus machines and any hopes of building a computer industry, the Americans allowed development to flourish and on February 16, 1946, launched the Open Source movement, although no one knew it at the time. On that day, the War Department's Bureau of Public Relations miraculously put out a press release (you can find it at the end of the book) titled "Physical Aspects, Operations of ENIAC Are Described." You could probably build an ENIAC today from reading the description, but a few more details were needed back in 1946 to construct a computer. So the Moore School, even more miraculously, ran a series of 48 seminars that summer on lessons learned, mistakes and all, from the ENIAC and the new design for the EDVAC—which wasn't completed until 1952.

Theory and Techniques for Design of Electronic Digital Computers, better known as the Moore School Lectures, drew 28 attendees from 20 universities, government research labs and companies. You can imagine that every single one of them returned home and insisted to their bosses that they too should be building digital computers. Shared ideas. What a concept. It set off a competition for the best and the fastest computers that still rages today. Meanwhile, the British computer industry smoldered in the embers of the burnt Colossus machines. What a turning point. The British Empire had died and now was officially cremated.

Work on computers and their acronymic names took off

everywhere, with added urgency when the Soviets tested their atomic bomb in 1949: Turing's MADAM; Maurice Vincent Wilkes's EDSAC—Electronic Delay Storage Automatic Calculator at Cambridge; Rand Corporation's UNIVAC—Universal Automatic Computer; and topping them all acronymically, Nick Metropolis's MANIAC—Mathematical and Numerical Integrator and Computer. Each contained more vacuum tubes, more memory and more sophisticated programming. Most were funded directly or indirectly by the Army or Navy, to research guided missiles, hydrogen bombs and aircraft design.

These were still vertically organized companies and institutions that insisted on their own computer structure and their own set of codes and software. That wouldn't change until 25 years later, when Ted Hoff created the microprocessor and accidentally broke the barrier between design and manufacture. Von Neumann's work had established a computing architecture. But the next move, like all the improvements Watt made to his original steam engine, was to get the hot, power-hungry vacuum tubes out of the loop and make computers commercially viable. And slow relays or a wave sloshing around in mercury was no way to store data.

Transistors and Integrated Circuits Provide Scale

In the 1920s, vacuum tubes were taking off as amplifiers for radios. Tube manufacturers were multiplying like rabbits. Western Electric made tubes for AT&T, but so did RCA, Raytheon, Sylvania and even General Electric (it was, after all, just a light bulb with an extra plate). The stock market was hot for radio stocks, much as it would be for Internet companies 70 years later. Still, tubes were problematic; when they heated up, the materials inside broke down. Vacuum tubes would fail, early and often. This particularly annoyed the Army; it was investing heavily in tube-based radar and two-way radios. In 1926, to be a part of the mania, New Yorker Julius Edgar Lilienfield filed a patent: "Method and apparatus for controlling electric currents." It described a junction transistor that could replace tubes as amplifiers. He was a bit ahead of his time, despite describing a device even Ben Franklin could understand.

Certain materials, germanium for example, contain impurities which cause them to either carry around a lot of extra electrons, or lose a lot of electrons. Look germanium up on a Chemistry Periodic Table (sorry to unnecessarily return you to high school). It's right there in the middle, type IV. If it has phosphor in it, an element which has extra electrons, it is known as a negative-type, or N-type, material. Fair enough. Materials with boron or gallium impurities in them, on the other hand, naturally lack electrons. Inside this structure, there is what is known as holes, or regions that lack electrons. This is known as P-type material.

Sounds like a match made in heaven. If you connect an N-type and a P-type material, some of the electrons from the N will flow and fill holes in the P-type and create a region with no charge. Only a small region between the N and P will be neutral, sort of a demilitarized zone or junction. In other words, holes flow from P-type and neutralize some electrons. Ben Franklin, who got the direction of electricity flow wrong, is vindicated; he just didn't realize he was pointing out the direction of the flow of holes! No matter.

The P and N materials jammed together created a diode, same as the Fleming valve of 1904. In this PN diode, electricity only flowed in one direction, hence the diode or valve. Devices like this are also called semiconductors, sometimes they conduct, and sometimes they are insulators. So let's meet halfway, hence semiconductors.

Lilienfield figured out that, like the De Forest diode vacuum tube, if he could jam another material into the mix, NPN or PNP, he might be able to create the same amplification properties of the tube and make smaller radios that used less power and lasted longer.

Lilienfield's problem was that the device he described and patented was completely unbuildable at the time. The whole thing was a concept. I read his 1926 patent filing. The thing resembled what is now known as a field effect transistor, or FET. The middle section sort of floated above the other two, and the charge put on the middle section acted like a magnet, pulling in or repelling electrons. Lilienfield was brilliant. You can find a couple of million of these FETs in your PC.

And just because one couldn't be made didn't mean it wasn't needed. Take AT&T. It was using vacuum tubes to amplify long-distance signals. An amplifying box with lots of tubes might be placed every few miles, as a repeater. That's OK between New York and Philadelphia, but the line between St. Joseph, Missouri, and Sacramento, California, was a real pain. The tubes would burn out and someone would have to truck out into the middle of a desert or a mountain pass to fix the thing. Bring back the Pony Express! This was happening all the time, at random amplifiers.

At Bell Labs, back in 1936, William Shockley was assigned the task of finding a replacement for tubes. Within a couple of years, he figured out he needed either NPN or PNP transistors. But for a decade after he began the project, he couldn't for his life build one. Management was getting impatient as phone traffic was increasing rapidly after World War II ended. In 1945 they assigned two engineers to help him, John Bardeen and Walter Brattain.

Shockley had gone off to try to make field effect transistors like the one Lilienfeld had described, as well as a junction or bipolar transistor, which are better amplifiers than switches.

But Bardeen and Brattain needed something to build on so they created the point-contact transistor. Actually, they called it a transfer resistor, transistor for short. The clue was old crystal radios. On these prevacuum tube radios, a thin wire called a cat's whisker was moved up and down a crystal of lead sulfide or germanium until it tuned in a radio station. Bardeen and Brattain were intrigued by the properties of germanium because it fit the model of Shockley's concepts. They fashioned their own cat's whiskers out of phosphor-bronze wires. At first, they just connected them to N-type germanium, but not much happened. They then zapped them with a quick, high-current pulse, and basically fused them to the germanium. It worked because the high current diffused a bunch of the phosphor from the wire into the germanium, creating P-type regions in the germanium. This diffusion is known as doping. The rest of the germanium remained N-type, hence a PNP device, and one with wires already sticking out of it to boot! It first worked on December 23, 1947. AT&T didn't tell the rest of the world until June of 1948, buying time to file for patents as well as make more than one prototype.

Shockley was the boss, so he insisted that only his name go on the patent. Bell Labs lawyers got involved, and after doing a patent search, found the Lilienfield patents, so they hedged their bets. They applied for four patents. Bardeen and Brattain got their names on a point contact transistor patent and Shockley's name went on a junction transistor patent. Two different patents were filed on types of field effect transistors that Shockley conceptualized and B&B worked on.

They filed the patents just before the June 30, 1948, announcement. By November, the patent office had turned down the two field effect transistor patents because Lilienfield already had that invention, but the other two patents, 2,502,488 and 2,524,035 stood. AT&T got its amplifiers for long distance, the radio industry got receivers and amplifiers, and the computer business without quite realizing it, got the greatest Logic and Memory element it could have wished for. The three Bell Labbers would share the 1956 Nobel Prize in Physics.

Shockley got his junction transistor to work in July 1951. In September, Bell Labs held a Transistor Symposium. Because AT&T was considered a public utility, anyone could license either patent for a whopping $25,000.

Most of the vacuum tube manufacturers of the day

licensed the transistor patents. Texas Instruments, an oil well service company, applied for a license. So did a lot of firms that popped up overnight to license and exploit transistors, Transitron, Germanium Products, Clevite, and Radio Receptor. And eventually Shockley himself.

The transistor was an important step from mechanical to electronic. You just cooked some germanium and stuck some probes in it. That would turn out to be much cheaper than winding coils for relays or shoving filaments into vacuum tubes that could heat a small village when operating. Radios and computers quickly computerized. But in one of the strange twists to this story, AT&T would not use transistors in its phone network for another 10 years, as it was sitting on a decade worth of vacuum tube inventory.

Shockley correctly predicted a huge industry making transistors and semiconductors, and he considered himself a superstar. Having grown up in warm, sunny Palo Alto, California, he figured that being stuck in Murray Hill, New Jersey, where Bell Labs was located was no prize. So he quit in 1955 and headed west, but not home, thinking Palm Springs would be the perfect location. Nice weather, swimming pools, movie stars. He probably couldn't admit it, but in the back of his mind, he knew he needed lots of other smart people to make this huge industry and Bob Hope was just not the transistor type. So in February 1956, he set up Shockley Semiconductor in Palo Alto, near Stanford University, which his mother had attended. He recruited the best and the brightest out of research labs and universities to join him making these cool new devices (as opposed to hot vacuum tubes). Beckman Instruments funded him.

It didn't take long for Shockley to run the thing into the ground. Even though he had a transistor license, he spent his time perfecting the four-layer switching diode. You know, that's the device that's in every . . . , okay I can't think of anything that uses a four-layer switching diode. And as a manager, Shockley was awful. Everybody hated him. His team of Gordon Moore and Robert Noyce and Jean Hoerni had figured out how to make transistors cheaply and were working on multitransistor devices, but were restless.

In September 1957, eight employees up and left. They were known as the "Traitorous Eight": Gordon Moore, Sheldon Roberts, Eugene Kleiner, Robert Noyce, Victor Grinich, Julius Blank, Jean Hoerni and Jay Last (left to right from a famous photo of them). Someone's father knew some money folks in New York, who sent Art Rock to check out this group. He helped them cut a deal with Stephen Fairchild of Fairchild Instruments out on Long Island, New York, and cut himself into the action at newly formed Fairchild Semiconductor as well.

Within no time, they were selling transistors to IBM for $150 each. IBM had an insatiable appetite for semiconductors. There was a go-go economy in the 1950s and 1960s, big business was buying computers to automate accounting, and transistors beat vacuum tubes in cost, speed, power use and best of all, reliability.

Demand was so strong that most firms were selling whatever they could make. Germanium is a somewhat rare element, so tended to be expensive, while silicon is as abundant as the sand on the beach. Actually, sand on the beach is silicon. Silicon mixed with oxygen is glass. And silicon is right above germanium in the periodic table, so it has the same properties; it can be doped with boron to make it P-type and doped with chlorine to make it N-type.

Texas Instruments was doing well selling transistors, and

joined the rapidly growing military-industrial complex. It turned out that the equipment to log oil wells was useful as SONAR for locating submarines. Like everyone else, the company was making germanium transistors, and IBM was buying them as quickly as it could make them. In 1952, TI figured out how to make silicon transistors, which were cheaper to make.

Ed Tudor, the CEO of Indianapolis-based IDEA Corp, figured that with a Cold War going on, every American would want to put a battery-powered portable radio in their survival kit. In November 1954, using a design from Texas Instruments (and four TI transistors of course) IDEA came out with the first portable transistor radio, the Regency TR-1. Suggested retail price was $49.95, leather case and earplug extra.

In 1953, a Japanese tape recorder company, Tokyo Tsushin Kogyo, lobbied the Ministry of Trade and Industry, MITI, to license the transistor patent and the manufacturing process from Western Electric. With it, Sony, as the company would eventually be known, would shrink these radios down to shirt-pocket size. This act single-handedly pushed the Japanese into the semiconductor business, but it is telling that it took government intervention to get permission for the license. This industrial policy of the Japanese would backfire years later. More often, innovation in digital technology took place in Silicon Valley because that was the end market for computers and network equipment. Small at first, this Silicon Valley innovation would fly under the radar of Japan's MITI watchdogs, who often would allocate resources to companies to exploit these innovations years after they had played out.

In the 1950s transistors were discrete devices, at $100-plus a pop. In 1952, G. W. Dummer at the British Royal Radar Establishment published a paper describing a device with lots of components and no wires. He desperately needed such a device for more accurate radar, but no one could build him one. At TI, a scientist named Jack Kilby was given the task of packing a bunch of transistors together. Kilby put five transistors on a single half-inch-long piece of germanium.

Hold the champagne though. Kilby sort of cheated. He used tiny wires to connect the five devices to each other. His integrated circuit looked like a petrified centipede. Still, it was the first integrated circuit, and for it Kilby won the 2000 Nobel Prize in Physics. I met Kilby at an annual meeting for Texas Instruments shareholders in the 1980s. He was about six-foot-four, and had a giant head that you instantly realized must house a massive brain.

Meanwhile, further west, after reading TI's January 1959 announcement of its integrated circuit, Fairchild Semiconductor doubled its effort to get more than one transistor on a device. Jean Hoerni, who was a Swiss physicist, came up with what is known as the planar process. Hoerni used a process called optical lithography, which is similar to how photographs are made. He started with a piece of N-type germanium or silicon. Then he sprayed over it a material called photoresist. If you shined a light on the photoresist, it hardened, and then you could use a special chemical to remove the photoresist that was not hit by light. So Hoerni created a mask, with a bunch of openings where he wanted to diffuse in impurities, and then flipped on a light. Wherever the photoresist remained after the chemical bath, the impurities would not diffuse underneath.

So he started with this bulk N-type, opened a bunch of holes in it and diffused in impurities to create P-type regions. Done with that, he opened a bunch more holes in the middle of the P-type and diffused in impurities to create N-type regions. When finished, he had NPN devices, as many as he could fit. I think Hoerni started with eight devices.

But Hoerni did no better than Kilby. He still needed wires

to connect the devices. In early 1959, his colleague, Bob Noyce, came up with the solution. Noyce grew an insulator, Silicon Dioxide, which is glass, over the top of the entire circuit. Then again using a mask and photoresist, he cut holes in the glass where he needed to connect to the N, the P and the N regions. Noyce then deposited molten aluminum over the top of the glass, which ran into the holes to make a connection. One more mask and photoresist step removed unwanted aluminum so you were automatically left with flat "wires" connecting the transistors (versus Kilby's hand-installed wires). Ingenious and forward-looking. This planar process has been perfected many times and is still in use today.

Because of these simultaneous inventions, Texas Instruments and Fairchild would argue in court for 10 years over who invented the integrated circuit. They could have saved the legal fees because they ended up agreeing to joint ownership, Kilby inventing the integrated circuit and Noyce inventing the process to wire up the components.

So with eight transistors on an integrated circuit, or chip, Fairchild could go out to IBM and others and charge a premium price. But inevitably, customers would come back and say, "You know, I really could use 14 transistors on a chip, with these five connected to these others."

Fairchild was more than happy to oblige, at a price. But the larger the chips got, the harder they were to make. So instead of making larger chips with more similar-sized transistors, Fairchild worked on fitting a few additional smaller transistors on the same size chip. Bing, bang, boom—the learning curve was invented. Shrink, integrate, shrink, integrate. The learning curve IS elasticity. This was a Field of Dreams. If you made it cheaper by half, they would not only come, they would use three times as many. Just like comfortable cloth off a steam engine run Spinning Mule and Power Loom.

The transistor I described Hoerni and Noyce making was a bipolar junction transistor. It was a great device for amplifiers and switching high voltages. The point-contact transistor was better at low voltage switching, but could never be manufactured in volume, as it was too complicated. The field effect transistor, which Lilienfield described and Shockley built, was also great at low voltage switching, and this was the type the computer industry was dying for. The industry was already making single transistors. But making integrated circuits with these FETs was difficult. In 1962, Steven Hofstein and Fredric Heiman at the RCA Labs in Princeton, New Jersey, came up with a new type of FET, the metal-oxide-silicon field effect transistor, or MOSFET. It was magical in that the middle device, called the gate, of say an NPN, never actually touched the other two, known as the source and the drain. It floated above them on a sliver of glass, but still operated as a switch. A positive charge on the gate would attract electrons into a channel, and current would flow from source to drain. Remove the charge, and the current stopped flowing. Hook up a few of these MOSFET transistors the right way, and you've got an AND-logic gate. Remember, with the AND-logic gate, you get both Logic and Memory and can create any type of computer you can imagine.

By 1965, Gordon Moore, who ran R&D at Fairchild, had seen enough to say to a magazine reporter that the number of transistors per device would double every year for the next 10 years. It was not that bold of a prediction, he had already seen it in action. In 1975, he adjusted what is known as Moore's Law, to say that the number of transistors on a chip would double every 18 months. What he left unsaid was that costs per transistor would also drop 30% per year.

• • •

The computer industry was booming. IBM came out with bigger and faster computers every year, and charged its customers huge markups for the right to buy them for millions of dollars. No businesses complained; they were saving multiples of that automating their back offices. Digital Equipment Corp. in Massachusetts, for example, brought out a minicomputer for $150,000, the poor man's computer that would start the march to ever lower prices.

Still, these were big, ugly machines. The transistors or the integrated circuits that contained 10, then 100, then 1,000 transistors were crude building blocks. The magic was how you hooked the chips together to create your central processing unit. Memory was even cruder. Back in the 1960s, memory for computers was often ferrite cores, meaning wires wrapped around little magnets that would retain their polarity after being written to. The materials were cheap enough, but a lot of the expense went into labor and machinery to painstakingly wrap the wires around the magnets. Some automation helped, but 20 billion to 30 billion of these core memory bits were made each year. One statistic I read pointed to 30,000 computers in the world in 1968, so that is about a megabit per computer, sounds right. Computer memory sold for about 20 cents a bit, which was a big business.

That spelled opportunity. With two transistors, you could create AND-logic gates. With four or six AND-logic gates, you could create a flip-flop, or a static memory device that allowed you to set it to a 1 or a 0. And it would stay that way until someone came along and changed it. Clever design techniques got another type of memory bit, dynamic memory, down to just two and even one transistor. But at $100 per discrete transistor, a bit of memory still cost a heck of a lot more than the 20 cents per bit of ferrite cores. But Moore's Law just required a little time and a little patience and cost would become irrelevant. It would take until 1974 before semiconductor memory would cross the price per bit of ferrite cores and send them to the trash heap of history.

Moore saw this coming. So did Noyce. In 1968, they started a company of their own to exploit it, Intel, which stood for Integrated Electronics. They actually had to buy the name from a motel chain. Hotel, motel, intel, I don't even want to know! Art Rock and friends ponied up $2.5 million in two days. Moore and Noyce quickly brought over a young Hungarian from Fairchild, a chemical engineer named Andy Grove.

Their first product was a 64-bit static memory chip, the 3101. It sold OK, enough to keep the company going, shrinking and integrating newer chips. By 1971, Intel came out with the 1103, a 1,024-bit or 1K dynamic memory chip. Remember, it took three or four times less transistors for dynamic memory than static. The 1103 proved to be just what the market needed and sold like hotcakes. It was the best selling chip in 1971. By 1974, memory was a penny a bit. Intel was selling 4K memory chips for $40, and even though ferrite core got cheaper, it couldn't win the race down the cost curve.

Intel was getting pretty good at integrating transistors into these chips. In 1970, a Japanese calculator company named Busicom showed up and asked Intel to create 12 custom integrated circuits for a newfangled calculator. Intel assigned engineer Ted Hoff to the task, but told him that there was no way Intel could afford the design of 12 chips. Not wanting to spend his life designing these one-offs for every Tom, Dick and Harry–san that came along, Hoff suggested a special-purpose processor that he could tweak or reprogram every time a new, finicky customer came along. Management liked the idea and Busicom was willing to pay for the project, without asking for ownership of the processor. Dumb move by Busicom. Intel would go on to dominate the microprocessor business.

Mark the date: November 15, 1970. This was a major inflection point for the industry. The semiconductor industry, as nascent as it was, subtly went from selling piece parts or basic building blocks to selling designs, real intellectual property. It's just that no one at the time understood the significance.

Hoff and Intel had just busted the computer world wide open. Sure, all they had done was create a computer on a chip. Hey, Intel was a chip company, what did you expect them to do? But it was a programmable chip, meaning someone else could add value to the chip just by twiddling with some bits, by changing the list of instructions this computer on a chip would execute. By programming it. Little Billy Gates would figure this out by the end of the decade and he and Intel would go on to dominate the world of desktop computing.

This was the first major step in creating a horizontal business, where different companies have mutual incentives to create intellectual property with which they could all prosper, TOGETHER. The chip was worthless without a program. The program was just a bunch of fuzzy 1s and 0s without the chip to execute it. Like cotton without a textile mill.

Later, U.S. chip companies would figure out that they could just design the chips, and let someone else in Taiwan manufacture them, fracturing the industry into even more horizontal layers.

Hoff and Federico Faggin used 2,300 of those integrated transistors to come up with the 4004 microprocessor, using 4-bit-wide data paths. In other words, it did binary math with numbers 0 to 15: Four bits can only represent 16 numbers. It actually was part of a 4-chip set, that included a 256-byte ROM, and 32-bit RAM. It created numbers bigger than 16 by putting a bunch of 4-bit numbers together. Eight bits gets you to 256; 16 bits gets you to 65,536; 20 bits gets you over a million. The 4004 had a 10-bit shift register for multiply and divide, and executed a whopping 60,000 operations per second.

The folks at IBM, if they even bothered studying the 4004, would have surely laughed. IBM was charging millions for 32-bit computers, with millions of bits of memory. But what Hoff and Faggin implemented was a computer architecture that was manufactured using a technology that got cheaper by 30% every year. Like core memory, it could eventually intersect the cost and performance of those big honker machines. Not could, would.

Intel never rested. In November 1972, it came out with the 8008 with 3,300 transistors. It used an 8-bit data path instead of 4-bit to handle larger numbers, but still was mainly a chip for calculators. About the same time, Texas Instruments was creating its own microprocessors for its internal line of calculators. It decided not to sell processors on the open market so no one could undercut them on calculators.

The folks who should have made an immediate beeline to Santa Clara, California, were the humps at the Pentagon. These computers on a chip were the solution to their tonnage problem. Yawata-like precision problems still plagued the war effort in Vietnam. Tonnage trumped targeting. One precision bomb would have replaced a dozen bomber sorties.

In April 1974, Intel came out with the 8080, with 4,500 transistors that could do 200,000 operations per second. Matched with 1K or 4K memory chips and you've got yourself a real computer. OK, it may have been a chintzy little do-nothing machine, but it was a computer. Altair and Imsai were two early hobbyist computer companies, which used the 8080 and some memory that was loaded with switches on the outside of a box. Results of programs were written to little lights above the switches. If you could afford one, you added a "TV typewriter" and it was a personal computer. Bill Gates and Paul Allen wrote a BASIC computer language interpreter to write simple programs for this machine.

But Intel's competition came fast and furious. RCA was making an 1801 microprocessor that the automobile industry was using for engine control. As a kid in high school, around 1974, I used to sneak into an RCA facility and steal these 1801s and other chips in an attempt to make a home computer. I got a few lights to flash, but no computer! Mostek had the 6502, Motorola the 6800. Zilog mimicked the 8080 with its Z80, using the same instruction set inside the microprocessor, and adding a few extensions of its own.

With the help of two much smarter high school friends, Dave Bell and Greg Efland, I actually built a Z80 home computer, complete with 64 1K-memory chips, a graphics interface to a TV (my dad dug through the TV set so I wouldn't electrocute myself) and a Panasonic cassette recorder to store programs. We stole a copy of the Gates BASIC program and were on our way. Of course, we spent most of the time writing computer games, but what do you expect for a bunch of kids. I wrote the world's greatest clone of the arcade Dominoes game, honk if you remember it. We thought of starting a company, but noticed ads for computer boards from a company named Apple in computer magazines and decided, or our parents did, that college might be a better use of time.

I met Ted Hoff in the late 1980s. He told me a story

about how he would take his TV to be repaired, and the guys at the shop would tell him there was a problem with the microprocessor in the set and then ask him why he was laughing. OK, techies do have a strange sense of humor.

Of course, in the 1970s, Motorola was making TVs. Baby boomers were inundated with TV commercials that blasted "Quasar, dum-dum-dum, by Motorola." (Dumb was right; the Japanese soon dominated the analog TV business.) Intel watched many other companies do well with integrated electronics, some even provided by Intel. Why shouldn't they slap a battery or a power supply and a little plastic around their parts and make even more money?

Digital watches were hot in the early 1970s so Intel bought a company named Microma, in 1973. The average digital watch with a cool red LED display sold for more than $100. But as volume increased, Intel was able to make the integrated circuits inside the watches much cheaper. Even though it was driving costs down the learning curve, no one quite understood the impact of lower costs on the end markets. Microma was stuck with inventory of expensive watches that wouldn't sell, as cheaper models were introduced. As others sold the necessary chips to new watchmakers, the market was flooded with cheap watches. Intel took a bath and quietly sold Microma in 1978, swearing never to forget this valuable lesson. I once asked Gordon Moore about the whole Microma experience, and he quickly pointed to a Microma watch on his wrist and told me he wears it often to remind himself to never be that stupid again. Intel's lesson? Make the intellectual property, not the end product.

The same thing happened with calculators. Texas Instruments and Hewlett Packard had a great business selling calculators for more than $100 each in the 1970s only to see prices drop to $20 as the cost of the chips inside raced down the learning curve.

Software and Networks

In 1964, a consortium of MIT, Bell Labs and General Electric (which made computers for a while) announced a system called Multics, which was a time-sharing computer with users connected via a video display terminal and an electric typewriter. A green screen beat punched cards, but computers were still mainly used for number crunching tasks. In 1969, Bell Labbers Ken Thompson and Dennis Ritchie, with beards a la ZZ Top, fixed the then dead Multics software. They wrote a simpler version for Digital Equipment Corporation's PDP minicomputers and named it UNICS, which stood for "castrated Multics." This is the type of joke that programmers live for at 4 A.M. while waiting for their programs to compile. The name was quickly changed to UNIX.

On December 9, 1968, a symposium was held in San Francisco called the American Federation of Information Processing Societies' Fall Joint Computer Conference. With a billing like that, I'll bet you're sorry you missed it. A group of researchers at the Stanford Research Institute led by Doug Engelbart had been working since 1962 on a topic they called "Augmented Human Intellect," and were set to present their findings. A snoozer right? It would turn out to be one of the most important gatherings in technological history.

Engelbart got up and demonstrated a few things his team

was playing with, built onto something called NLS, or oN Line System. On a computer screen with both graphics and text were multiple windows, a text editor with cut and paste, and an outline processor. A "mouse" controlled an on screen pointer as a cursor. Multiple users could connect remotely—in fact the demo was connected live to Menlo Park, 45 miles to the south. There was hypertext to be able to "link" to information anywhere on the computer or network. If you were stuck, a help system would provide assistance based on the context of what you were looking for. The 1,000-plus attendees were stunned. This was 1968, with hippies roaming aimlessly through San Francisco. Yet here was this guy who laid out the map for the personal computer industry and Internet that would unfold over the next 35-plus years. And the microprocessor was still two years away from being invented!

Doug Engelbart is the unsung hero of the computer and information revolution. And I'm not just saying that because he lives next door to me and doesn't call the police when my kids are outside screaming so loud that sometimes I feel like calling in the National Guard to quiet them down.

Engelbart's augmentation ideas, while not implemented for another 15 years, took the horizontalization of the technology business a step further. Several different companies could take a chip, program it with an operating system, add applications, user interfaces, networking, databases and create a system that could, well, augment humans. It was the augmentation part that IS the value of computers. What do I mean by this? I once interviewed for a job with United Airlines in Chicago (truth be told, they sent me a first-class ticket at the precise time I was trying to figure out how to visit my long-distance girlfriend, now wife, who lived in Chicago). I got a tour of their facilities and walked through room after room filled with scores of people sorting airline tickets. Wall Street had this same back office paper mess back in the early 1970s. This was the early or mid-1980s and even I knew what a waste of a task that was, that computers could either read the tickets or a database could track them and eliminate the sorting mess. I still cringe at the thought of all that sorting. None of the sorters seemed all that thrilled, except they were fully employed. I was worried one was going to throw a wooden shoe at me!

I politely declined the company's offer.

At a certain cost of computing power, when elasticity kicked in, computers had to be cheaper than rooms of humans. Augmenting often meant replacing, and helping those who remained. It also meant the creation of jobs elsewhere to create these tools.

Bill Gates and Steve Jobs and Michael Dell and Lotus spreadsheet founder Mitch Kapor—these guys are great implementers. More power and riches to them, of course, but Doug Engelbart laid out their future and that of the personal computer. Not much more needs to be said (that others haven't already overwritten).

An entire industry followed Doug's map to create high-margin intellectual property. Computers were not just for boring accounting functions; they really could augment the human race, and increase efficiency and productivity by replacing costly repetitive human functions.

From the 1978 introduction of the Apple II computer to the 1981 announcement of the IBM PC (IBM bean counters estimated they would sell 250,000 over the life of the PC), the world has been flooded with smaller, cheaper and faster computers: 100-plus million new ones get sold every year. But these are no islands—the power of the computer is in its ability to communicate.

• • •

Let's go back to Bell Labs. In September of 1940, George Stibitz, the guy who created the Complex Number Calculator, was working at Dartmouth College in New Hampshire when he connected a Teletype (a 1930s invention that allowed telegraphs to be read as text) to a phone line and controlled his CNC at a Bell Lab in New York City.

Thus began the use of the telephone network, which is optimized for gabbing with Mom, as the medium for computer communications. No one thought this out, it just happened that phone lines were already running everywhere, so as computers were placed in the same offices everywhere, they used the phone network to communicate.

The problem was that from the very beginning, the phone company had cut corners. The human ear can hear sounds from a few cycles per second or hertz, to 20,000 hertz. Dogs can hear higher frequencies. But a voice conversation mostly occurs at below 3,000 hertz. So Bell engineers just put in filters to cut off frequencies above 3,000 hertz. They figured correctly that humans didn't need it to hold a decent conversation.

But advanced computers would. Data will gladly gobble as much frequency as it can get. Computer communications have been held back by that 3,000-hertz decision ever since. In 1962, Bell Labs invented the Bell 103 Modem, modulator-demodulator. Placed on each side of a telephone line it would connect a Teletype to a computer at 300 baud, or bits, per second. It used a technique called Frequency Shift Keying. Modems are two-way, so they originate and answer. A "1" is 1,270 hertz when originated; answering back, a "1" is 2,225 hertz. An originated "0" is 1,070 hertz and answered "0" 2,025 hertz. Pretty simple, right? Just a bunch of filters to figure out the frequency and 300 bits per second was no problem.

Because back in 1962 Bell Labs had no idea what types of phone lines were out there, it made the 103 modem industrial strength and designed it to work on almost any line. The company boasted it could communicate down barbed wire.

But as soon as you went above 300 bits per second, the whole thing broke down. And 3,000 hertz of bandwidth on phone lines is just not enough. So Bell Labs went back at it and played around with other techniques. The 1,200-bit-per-second Bell 212a in the mid-1970s used a technique called Differential Phase Shift Keying. Two integrated circuits designed by George Kutska and Hank Goldenberg at Bell Labs did all the encoding. Hank's chip had a bug in it that would occasionally insert a "{" in transmissions. This was no problem if you were sending text, you could just ignore it, but it reeked havoc if you tried to use the 212a for graphics. Other companies were allowed to buy the same chips, or copy the designs, and you always could tell if they didn't bother fixing the bug, as the modems would spit out intermittent left squiggle brackets. George (who became my boss when I worked at Bell Labs) and I used to play pranks on Hank, like forwarding every phone in the area to Hank's desk on the morning he returned from a two-week vacation.

For the longest time, use of modems was restricted on phone networks. Each state's public utility commission would set tariffs or rates for using data communications services. The modems that my group at Bell Labs designed were manufactured in Montgomery, Illinois, just outside of Aurora, home to Wayne's World. I was given the task of setting up some programs to test these modems. No problem. I would write the tests at Bell Labs in New Jersey and then use a modem to transmit the tests to the Western Electric factory in Illinois. This was 1981 and it should not have been a problem. But the Illinois Public Utility Commission had yet to tariff the 1,200-bit-per-second data service, so the factory couldn't use it, even though it turned out a thousand modems a day. Screwy. I learned a lesson about regulation, which is that it is never understandable and almost always wrong.

As the phone network shifted to digital switching in the 1980s, you would think data communications would have gotten easier. Nope. Same dumb Bellheads assumed their network would only be used for voice calls. The first thing a digital phone network does is convert your analog voice into digital bits, using a technique called pulse code modulation. It samples your voice 8,000 times a second and converts it into a 7-bit symbol, representing 1 of 128 possible amplitudes or signal strengths. The eighth bit is used to control the switch. At the other end, it is relatively easy to reconstruct your voice. Seven times 8,000 equals 56,000, which of course is the highest rate a modem will work over the phone network today, because of these old compromises. Notice I didn't say the highest rate over a phone line, but instead through the entire network. New techniques like Digital Subscriber Line can send a million bits per second, some even 45 million bits per second over the phone line between your home and the phone company's central office. Notice I said "can" not "will"; the phone company still controls those lines and continues with its dated Bellhead view of data. Something was needed to replace those copper wires.

Let's go back in time again, about 100 years. At the time that Swan and Edison were inventing the filament light bulb, others were still working on more conventional means of home lighting. An inventor in Concord, Massachusetts, named William Wheeler was convinced he could convert the electric arc lamp that worked out in the streets for home use without fire danger, by installing an arc lamp in the basement of a house, and then running metallic coated glass pipes to rooms in the house, like a giant mirror tube. He patented his idea in 1881. Of course, it was a failure, because each reflection off the mirror would degrade the light. But he did quite accidentally invent the fiber-optic cable, only no one cared.

Around the turn of the century a Brit named Charles Vernon Boys came up with a way to pull extremely thin cables, called drawn quartz, on which he hoped to transmit light. Didn't happen, but his fascination with growing and pulling led him to write a not so blockbuster book in 1937 titled *Weeds, Weeds, Weeds.*

That reminds me of how one invests in new technologies. You plant lots of seeds and hope to get lots of flowers and a few trees, maybe even some towering redwoods. You never know which technologies or which companies are really going to work. More often than not, you get *Weeds, Weeds, Weeds.*

Electrical signals were great for carrying information, voice and data over phone lines. But radio waves were even better because they allowed for much higher frequencies, so radio and eventually television shows would be broadcast using waves in the air. Radio technology, especially for radar, improved during World War II. Using this technology, in 1947, AT&T began replacing long-distance lines with microwave towers. A few birds got fried along the way, but who would miss some smelly pigeons? AT&T found it could cram hundreds of phone calls onto these microwave signals and send them 30 or so miles, before the earth's curvature came into play. You still see microwave towers sprouting out of phone company buildings all over the country.

But light was the ultimate carrier of information. A picture is worth a thousand words. There was a need for a light source that could work in a tight frequency range, travel great distances, and be controllable, meaning you could turn it on and off rapidly. You could shine a flashlight in one end of the fiber and turn it on and off and if you squinted, you might be able to make that out at the other end, but that would be a pretty flimsy modem.

In 1917, no less than Albert Einstein (you knew he would be involved in this story somehow) theorized a photoelectric effect. If you shone light on a metal, it would emit electrons. But no one could adequately explain why. Isaac Newton had thought light was made up of particles, but in the early 1800s, it was Thomas Young who set up an ingenious experiment. He shined a light source through a board with a double slit in it, and noted the patterns on a piece of paper on the other side of the slits. The patterns looked similar to patterns made when you drop two rocks in a lake at the same time. At certain points, the waves cancel each other out. He concluded that light was made of waves. A particle would go through one slit or the other, but not both. Experimenters would shine light on metals and measure the energy of electrons emitted. To their surprise, they noted that as you increased the intensity of the wave, the energy in each electron did not increase. Einstein started thinking about it, and theorized that light was made, not of a continuous wave or particles, but instead, of both, bundles of waves he called photons. The higher the frequency, the higher the energy of the photons. Einstein received a Nobel in 1921 for explaining the photoelectric effect. The Nobel committee didn't quite get his relativity stuff.

But Einstein didn't quite nail the whole photon thing. An Indian physicist named Satyendra Nath Bose added to the theory in 1924, with a birds-of-a-feather-flock-together theory. Bose said that light operated in specific quanta or packets (these were Einstein photons) but that, like birds, if they were in the presence of other photons, they would tend to operate at the same frequency. Einstein thought this might also be true for atoms, and the name Bose-Einstein Statistics was applied to this phenomenon. This is the principle that lasers would eventually implement. The only flaw in the theory was that at very low temperatures it didn't work, as the quanta disappeared. A new name solves that, Bose-Einstein Condensation.

Charles Townes, a CalTech PhD, started working at Bell Labs in 1939, playing around with vacuum tubes and generating microwaves. As the war broke out, his boss came in and told him to work on radar bombing systems, using 10-centimeter waves. The bigger the waves, the bigger the antenna and there was only so much room on a plane. So they told him to try working with shorter wavelengths, down to 1.25 cm. He worried that water molecules would absorb microwaves at that short of a wavelength, rendering radar inoperable in fog, which of course defeats the purpose. He tested it out and sure enough, water absorbed these short wavelengths. Townes had inadvertently invented the microwave oven, ending forever the terrible problem of cold leftovers. But that's another story.

And that's how it goes at Bell Labs, you play around with one thing and invent another. But the task of Bell Labs was to improve communications. Even into the 1940s, communications was just two wires with electrical signals representing voices running down them. In 1941, a University of Michigan undergrad and MIT grad, Claude Shannon, joined Bell Labs. As per custom, no one told him what to do. So he started trying to apply mathematics to communications. He wasn't interested so much in the signals, but in the probability that what one shoved into one end of a channel would come out the other end intact, through all the noise that impairs the signal. And what if you shoved an encoder and decoder on either end. How much information could you transmit? Or how clear could the voice signal be?

Through a series of papers starting in 1948 came Shannon's Law, which calculated the maximum throughput of error-free information through a channel with a certain amount of noise in it. Put another way, Shannon laid the foundation of modern information theory, turned encryption into a science (the fallibility of Enigma-like machines is behind us) and set limits on what wireless networks could be used for. Pretty cool for 1948. Even today, an entrepreneur will occasionally introduce some newfangled communications system or protocol that promises massive throughput and it usually takes 24 hours for someone to punch holes in it for violating Shannon's Law. There is no perpetual motion machine, in real life or for digital communications.

While Shannon was working on his communication theory, Charles Townes had left for Columbia University in 1948, to return to pure science. There he connected with postgrad Arthur Schawlow. They attempted to build microwave devices that emitted waves at even shorter wavelengths, but kept getting stuck. Townes kept noodling on the 1.25 wavelength waves getting absorbed by water molecules and turned the whole thing around. Perhaps he could stimulate molecules to emit microwaves, like Einstein predicted. Schawlow finished his fellowship, married Townes's sister, and around 1951 took a job at Bell Labs. Can you imagine the Thanksgiving dinners with that family!

Townes used ammonia in a ruby cavity (is that why Dorothy's slippers sparkled?), which he pumped with energy and got the chromium molecules in the ruby to emit radiation. He called his invention the MASER, which stood for

Microwave Amplification by Stimulated Emission of Radiation, and patented it in 1953.

Visible light is an even shorter wavelength than microwave and the logical next step, but it wasn't so easy. Gordon Gould, a graduate student at Columbia, conceptualized the laser. He would even have his lab notebook with the word "laser" in it notarized in 1957. But he thought you needed a working model to get a patent. Meanwhile, Townes got a consulting gig at Bell Labs and started working with Schawlow—if you could mase microwaves, you could lase light. Schawlow came up with the idea of putting mirrors on either end of the cavity, at a precise distance, which was of course an exact multiple of the wavelength. This way, the emitted waves would reflect back and forth while being amplified until they shot out of a slit or half mirror at one end. And the light coming out was coherent, as Bose predicted, meaning it was one package of photons and one phase of light, not the scatter of light that comes out of a light bulb. Schawlow also suggested that they could get solid materials, instead of just gases and liquids, to emit waves.

In 1958, as Kilby was putting five transistors onto a thin piece of germanium, Townes and Schawlow released an article in the special holiday issue of *Physical Review*, describing the concept of a laser. And they applied for a patent. Gould filed for his patent in 1959, a little too late, even though no working laser existed. In early 1960, Townes and Schawlow got their patent, although Bell Labs actually owned the rights. In May, a scientist named Ted Maiman demonstrated the first laser at Hughes Aircraft in California. Maiman wrapped a flash lamp, like the one in your camera, around ruby in a cavity with mirrors on each end, just as Townes and Schawlow had described but hadn't yet built. With each flash of the lamp, the device would pump out coherent light, pink light actually, but only in pulses. The Bell Labs crew went back and constructed a continuous laser by using an arc lamp instead of a flash lamp. Those arc lamps just wouldn't die! As you can imagine, the lawyers got involved as to who really owned the rights to the laser and it wasn't until 1977 that Gould got a few patents, and Bell Labs kept the rest.

So Einstein's theory was real, and work on lasers took off. In 1962, a semiconductor injection laser was invented, which is exactly what would be needed 35 years later to get gigabit data networks to operate over fiber-optic cables.

What about those cables? Between the time Vernon Boys was pulling quartz cables and the early 1960s, most work on fiber-optic cables was focused on transferring images. Doctors wanted to stick thin glass rods into patients and look around. Some thought they could use fiber optics to shine light on a rotating drum to generate television images. American Optical in Southbridge, Massachusetts, offered various types of fiber-optic cable for these and other industrial uses. In 1961, American Optical scientist Eliot Snitzer proposed a form of cable so thin, light would zoom down it without reflection or interference, and act as a waveguide, a medium just the right size for a wave to travel along without interfering with itself.

In 1970, it all came together. Three researchers at Corning Glass in upstate New York—Robert Maurer, Donald Keck and Peter Schultz—patented a fiber made from fused silica and doped titanium that was so thin and so pure, light could travel for a thousand miles in it. Their patent, number 3,711,262 for Optical Waveguide Fibers, combined with semiconductor diode lasers, finally solved the limitations of copper wires.

Just as Tesla had once thought that flashes of electricity would make a great death ray, many now thought the same of lasers. Lasers would play a huge role with the military, but as an uninterruptible and portable communications medium. They could connect computers, communicate targets to riflemen or precision-guide weapons.

AT&T was thrilled to have fiber optics carry voice calls. By 1976, it had systems running at 6 megabits per second, which meant 2,000 phone calls could be carried on one fiber line. By 1977, it was installing 45-megabit fiber systems capable of 15,000 phone calls. An early problem was that lasers failed quickly. In 1977, Bell Labs invented a semiconductor diode laser that, they guessed, could last for a million hours or about 100 years. This was the reliability needed if you were going to drop them under the ocean for transoceanic fiber connections. If only I. K. Brunel was around to see it.

Fiber was great for handling phone calls. As voice conversations were point to point, two people wanting to talk to each other, you could now bundle a bunch of these calls onto fiber between Chicago and St. Louis.

People wanted their computers to talk to each other as well, to exchange files, to allow remote users to swap e-mail (we'll get to that soon), heck, even to share printers, all the things that Doug Engelbart—remember our unsung hero?—had demonstrated in 1968.

Modems at each machine could now dial each other up. Unix has a command built in named "cu" for call up, which did just that. Its modem dialed up another machine's modem whenever it wanted to send it a file or message. (Don't tell anyone, but George Kutska and I used to write programs to set off "cu" commands to Hank's house at 4 A.M.) But modems were slow. And a bunch of computers in the same office should have some way to communicate at high speeds other than trying to tunnel through a thin phone line.

● ● ●

In 1957, before the integrated circuit was born, the Russians shot a satellite into earth's orbit. Sputnik I scared the Powers That Be in the U.S., and the Cold War added a new dimension beyond nukes: The Space Race. In response, the Advanced Research Projects Agency was created (so was NASA) and, tellingly, located inside the Department of Defense. One of the issues of the day was the idea that a nuclear blast (I've learned never to trust anyone that pronounces it nu-cu-ler) would wipe out the phone network and all communications lines and disable the command and control structure of U.S. defense. The president could order a launch, but if no one could get the message, what would be the use?

In 1961, Leonard Kleinrock at MIT proposed a PhD thesis called "Information Flow in Large Communication Nets," and this provided the theory and proof for packet switching, although it wasn't called packet switching, not yet, and it was still a theory.

The North American Aerospace Defense Command, or NORAD, was in charge of early warning and control. It didn't want no stinking theories, it wanted something it could use. NORAD was nervous about being out of touch, especially with its command center dug into the mountains near Cheyenne.

So the Air Force sprinkled money around for research on ways to resolve the vulnerability of communications networks, which were dependant on centralized phone switches. Some of it went to the RAND Corporation, a Santa Monica, California, think tank spun out of Douglas Aircraft in 1948 to worry about such things. One researcher there, Paul Baran, was an electrical engineer who had worked at nearby Hughes Aircraft. In August 1964, he laid out his theory in a paper titled "On Distributed Computing." You can read it at RAND's website. Baran described standard message blocks and "store and forward" transmissions and hot potato routing. This was almost precisely what the Internet became, but he never used the word packet. It was Donald Watts Davies, a British mathematician working on block-switch networks in 1965 who came up with the name.

The best description is often attributed to Baran, but I don't think he ever said it, in fact I'm not sure who did (I got it, appropriately, off the Internet), but it is revealing:

> Packet switching is the breaking down of data into datagrams or packets that are labeled to indicate the origin and the destination of the information and the forwarding of these packets from one computer to another computer until the information arrives at its final destination computer. This was crucial to the realization of a computer network. If packets are lost at any given point, the message can be resent by the originator.

So there you have it. If you could install a computer at various points in the circuit-switched phone network, it would become a packet-switched network, and would withstand not only broken links, but a full-scale nu-cu-ler winter.

Still a theory, though. Larry Roberts at MIT proposed a collection of computers hooked together via packet switching, which turned into ARPANET. ARPA put out a request for proposals, the infamous RFP, for Interface Message Processors, or IMPs, which would be the store and forward computers to handle the hot potatoes, er, packets. I think that if the name hot potatoes had stuck, white papers describing how the Internet works would be a lot easier to understand. The ARPA RFP basically described a robust enough system that could suspiciously run Doug Engelbart's NLS operating system. Hmm, wouldn't be the first time government jobs were specified this way!

Bolt, Beranek and Newman, later known as BBN, got the contract from ARPA for ARPANET, using a Honeywell minicomputer for the IMP. The IMPs talked to each other using a Network Control Protocol. Hey, these were computer guys, they thought the ENIAC was a cool name.

In September of 1969, an IMP was set up at UCLA and in October, another was set up at Engelbart's office in the Stanford Research Institute. They were connected by a 50-kilobit-per-second connection that AT&T provided. One can imagine that AT&T was not at all enthusiastic about the project, since packet switching endangered the phone network. But there was probably pressure to act patriotically—plus the government paid good money.

Leonard Kleinrock, the MIT theorizer, of course joined the ARPANET project, since it was his theory being implemented. I sat next to him at a dinner in 2000, and he gladly recounted the story:

He was at UCLA and on the phone to Stanford.

"OK, we are about to send an 'L,' let me know when you see it," Kleinrock told the Stanford folks.

"There it is, we got an 'L' [sound of applause in the background]"

"OK, OK, just a second, hold on, we are going to send an 'O.'"

"[Screams in the background] Oh my God, we just got an 'O', keep going."

"Get ready, here comes a 'G,' let me know when you get it."

"Did you send it?"

"Yeah, we just sent the 'G,' did you get it?"

"Uh, the Honeywell just crashed."

"Who cares, we did it, success. Get out the champagne!"

• • •

How telling. There were two computers hooked together, yet they still tried to send a LOGIN request from one to the other. And the damn thing crashed. Sound familiar? More than three decades later it still happens to people like me several times a week. But the Internet was born. A Santa Barbara IMP went up in November and Utah in December 1969.

One of the researchers at the Stanford site, Norm Abramson, was a surfer dude who spent a lot of time in Hawaii. The University of Hawaii had locations scattered across the Islands and was trying to figure out how to hook up a data network among them. It couldn't afford to run undersea cable and modems were too slow, so they hit on the idea of using radio signals to transmit data. The problem was interference. Maui might transmit at the same time and step on the Big Island's signal. It could use packet networks, but that didn't solve the interference problem. In 1970, Abramson devised a system that checked for errors in the packets received. If the packets had errors, the receiver wouldn't send an acknowledgment signal back. If the sender didn't receive an acknowledgment (hence a collision or errors occurred), it would wait a little bit, actually a random period of time, and then resend the packet. Simple, yet effective, AlohaNet became the first Local Area Network, even though that local area spread across hundreds of miles.

By 1971, there were 15 nodes scattered across the U.S.: UCLA, Stanford Research Institute, UC Santa Barbara, University of Utah, BBN, MIT, RAND, SDC (which I think is the State Data Center, part of the Census Bureau), Harvard, MIT's Lincoln Labs, Stanford, University of Illinois in (scenic) Urbana-Champaign, Case Western Reserve, Carnegie Mellon, and NASA/Ames. This was a very interesting mix of academics, think tanks, government and quasi-military organizations. BBN was the only corporation, but of course, was getting ARPA funding to run the network. Since 14 other sites were a lot to keep in touch with, in 1971 Ray Tomlinson at BBN wrote a message reader and writer so BBN could send and receive notes on the system. He used @, the "at" sign, to denote the destination. E-mail was born.

One of the ARPANET researchers was in charge of giving a demonstration to some bigwigs from AT&T. Keep in mind that the phone network was engineered to fail for only 2 minutes every 40 years. That is one of those 5 nines to the right of the decimal point or 99.99999% reliability. As the story goes, Bob Metcalfe was in the middle of demonstrating the packet network when, like any good demo, it crashed. This put smiles on the face of those 10 AT&T execu-humps, and they merrily skipped back to HQ singing the stillbirth of packet switching. Of course, they were right for another 30 years, but packet switching would eventually be trouble for circuit-switched phone networks.

With the success of its network, ARPA became DARPA, to remind everyone it was "Your Defense Dollars At Work." The Network Control Protocol was OK for 15 nodes, but something new was needed for larger networks. Bob Kahn and Vint Cerf worked out the future and published a paper called "A Protocol for Packet Network Intercommunication," which described, TCP, or Transmission Control Program. Coupled with IP or Internet Protocol, TCP/IP has been the backbone of the Internet ever since.

Metcalfe actually was a member of the research staff at MIT and a PhD student at Harvard working on ARPA contracts. He took a job at the Xerox Palo Alto Research Center in June of 1972 while still finishing his thesis. Not surprisingly, the thesis topic was "Packet Communications."

Metcalfe was playing around with a bunch of new Alto workstations, and trying to devise a fast network to hook them together and to laser printers that Xerox was hoping to sell in large numbers. This was around 1973, and Xerox PARC was still a playground for dreamy scientists. Conference rooms didn't have tables and chairs—they had beanbag chairs.

But Metcalfe was no dreamer. He had read a paper on Abramson's AlohaNet and liked the idea. But what works for radio signals across islands might not work for busy computers connected with cables. The problem was AlohaNet's random retransmission, the random amount of time before you resend the packet. You could never get much throughput on a shared network if each machine just made up the amount of time to wait.

Years later, I asked Bob about this and he told me:

> The real problem with AlohaNet was that at higher loads (above 17%) retransmissions would too often re-collide bringing down the network. Ethernet adjusts its retransmission intervals based on an estimate of traffic backing off. The estimate is made by counting the number of retransmissions required to get a packet through, each retransmission implicitly increasing the traffic estimate and therefore doubling the mean of the retransmission interval.

So Metcalfe, working with David Boggs, came up with this new scheme. First, each computer would check the network first and "listen" for a carrier, meaning someone else was transmitting a packet. If it heard one, it would wait to transmit its own packet. Aloha didn't do this. In this new scheme, collisions would only occur when two computers "listening" heard silence, and decided at the same time to transmit their packets. Second, the transmitter wouldn't wait a random amount of time if a carrier was heard or a collision occurred, but instead, would first check how much traffic there was on the network. If there was only a little traffic, it would wait a random but short amount of time. If traffic was heavy, it would wait a random but longer amount of time to resend.

Metcalfe actually built one. It could transmit 2.94 megabits per second over coax cable, which, Bob told me, "is what you get by using the Alto's 170 nanosecond system clock, ticking twice per bit."

They nicknamed the two original workstations Michelson and Morley, after two turn-of-the-century scientists who refuted the theory that the universe was filled with some mysterious lumeniferous (light carrying) ether. In a memo on May 22, 1973, Metcalfe described his network and called it the Alto Aloha Network. He later changed it to EtherNet, then Ethernet, then The Xerox Wire, and then Ethernet again and then the 802.3 CSMA/CD LAN and then fortunately, Ethernet again.

It wasn't known outside of PARC until July 1976 when Metcalfe and Boggs published a paper in the *Communications of the Association of Computer Machinery* magazine, titled "Ethernet: Distributed Packet Switching for Local Computer Networks." The name Ethernet stuck. In the paper, they described their CSMA/CD scheme, Carrier Sense Multiple Access/Collision Detection. Local area networks were born. In 1977, Xerox received patent 4,063,220, "Multipoint data communications system with collision detection."

In 1979, Intel and Digital Equipment Corporation quickly signed onto the Ethernet standard that operated at 10 megabits per second. But as Xerox was never good at commercializing computer products, Metcalfe left to start his own company to commercialize Ethernet. He called it Computers, Communications and Compatibility or 3Com.

Well, not everyone wanted to be compatible. IBM went it alone, inventing something called Token Ring. Each computer was hooked to a cable connected in a ring. Tokens would be passed around, and if you wanted to send a packet, you had to grab an unused token, claim it as your own, and then send the packet. The receiver would read the packet and then "free" the token for someone else to use. Token Ring could be run much faster than Ethernet, at the time anyway. But the chaotic nature of Ethernet would prove over time to scale to faster speeds, and as no one was in control, the decentralized approach won out.

I got a tour of PARC in the early 1990s. Management touted how they had modernized, that Xerox could now commercialize projects, blah, blah, blah. When I asked about the beanbag chairs, they sniffed that that was the old Xerox PARC, that now they were serious. In 1999, I was asked to meet an entrepreneur at Xerox PARC, who was trying to commercialize a virtual whiteboard collaborative thingy, I never really did understand its uses. We met in a small conference room, table and chairs, and I was terribly disappointed. But sure enough, on the way out, we walked past this huge room filled with dreamy scientists sipping lattes, each resting comfortably in red, green and blue beanbag chairs. It's good to know that some things never change.

Packet switching, or connectionless connections, is more expensive than a point-to-point circuit, at first. But over time, the shrink and integrate learning curve kicks in, and the cost of packet switching plummets and the benefits swamp the old way of doing things. Finally, in 2001, the business of switching telephone calls died. Worldcom and Global Crossing tried to revive the corpse with accounting tricks to buy themselves time to move to packets, hence the accounting scandals in 2002.

In the fall of that year, I sat at the same table with Bob Metcalfe during a presentation on the future of networking. The talk was dull, but as a slide demonstrating uses of 10-gigabit and soon 100-gigabit Ethernet networks flashed on the screen, I noticed a funny look on Bob Metcalfe's face: He was half stunned, half amused that Ethernet is still scaling from those 2.94-megabit-per-second beginnings.

Metcalfe is also known for another famous observation—that the value of a network goes up by the square of the number of nodes attached to it. The official definition of Metcalfe's Law is: The value of a network grows as the square of the number of its users.

Actually, for those keeping track at home, it's $n*(n-1)$. A single node has no connections, two nodes have 2 connections, one in either direction, three nodes have 6, etc. This becomes the scale of the Web when millions of nodes are connected, and Metcalfe's Law is what made Napster and peer-to-peer file sharing such a huge phenomenon.

Bob told me, "Unlike Moore's Law, Metcalfe's Law has never been actually numerically true. It's a vision thing. You can quote me."

Local area networks, LANs, became extremely popular in hooking up mainframes and minicomputers. But it was the personal computer boom of the 1980s that dragged LANs into the mainstream. And they were not all the same flavor. There was Ethernet, of course, but there was also Hyperchannel, Arcnet, Omninet, Wangnet, Token Bus, Token Ring and many others. Novell had a networking protocol that ran on top of Ethernet called Netware. But so did UNIX and TCP/IP and IBM's SNA and Digital's DECNET. What a mess.

Around 1980, Xerox PARC gave Stanford University a bunch of Alto workstations and some of its new Ethernet networking cards. Like any large university, Stanford had a huge collection of computers: mainframes, minicomputers, and even some homegrown Motorola microprocessor-based machines built at Stanford by grad student Andy Bechtolsheim, who would later use them at Sun Microsystems. Plus, all these computers were scattered among the different schools doing research—the med school, the engineering department, the business school, etc.

Len Bosack in the computer science department, and Sandy Lerner in the B-school get credit for inventing routers (multiprotocol routers actually, the IMP was a single-protocol router). The guy in charge of all of Stanford's computers, Ralph Gorin, was interested in hooking them together, not the easiest of tasks. But, hey, this was Silicon Valley and Stanford had lots of enterprising people, so a bunch of them created a box that could connect these machines, even if they each used different protocols for networking. The "Blue Box" they came up with (it was in a blue box) was the first multiprotocol router. Like the Interface Message Processor that implemented the original ARPANET, the box was a store and forward device. It looked at the header of each packet, regardless of protocol, and decided where to forward it. But you could own one yourself, inside your own location, rather than out in the Internet.

The Blue Box was a combination of hardware and software. The hardware, much of it designed by Bosack, had to be fast enough to deal with these packets flying around, but the secret sauce was the software. William Yeager at the medical school wrote much of the original code.

A couple dozen of these routers were set up around Stanford. Other universities, hearing about it, wanted routers too. To sell them, Bosack and Lerner set up cisco (as in San Fran . . . it had a small c for the longest time). That was

December of 1984 and like all good Silicon Valley legends, it began in their living room on Oak Grove in Atherton. I mention the location because I drive by it every day—I live around the corner. Bosack and Lerner were assembling routers at home and selling them as fast as they could make them.

In April of 1987, Stanford licensed both the software and board designs to cisco in exchange for $19,300 in cash and $150,000 in future royalties. But cisco was on its way, backed by venture capitalist Don Valentine, who had backed Apple. By 1990, Bosack and Lerner were out of cisco and a new CEO, John Morgridge, took over.

Corporate America now had local area networks and routers to connect them to each other and to the ARPANET, which was rapidly evolving into the Internet. Most packets stayed local, on LANs. Files were transferred between computers, or from computers to local printers. It was the 80-20 rule, 80% of traffic stayed local, and 20% had to go through a router to interconnect to another network. These same routers were also critical to the growth of the Internet.

Not only did routers hook LANs to wide area networks, or WANs, that made up the Internet, but increasingly, cisco routers became the Internet's backbone. New companies like UUNET and America OnLine would use cisco routers in the middle of their networks to move packets around, as well as at the edge of their networks to connect to banks of dialup modems so users could call in and connect.

In 1991, a physicist at the Particle Physics Institute CERN in Geneva, Tim Berners-Lee, was tired of the hassle involved in sharing research among scientists. Leveraging the hypertext Doug Engelbart demonstrated 23 years earlier, he wrote some code-creating links across computer networks, and called it the World Wide Web. He didn't patent it, didn't start a company, and didn't get royalties. But he did get knighted and so received the title the Knight Commander of the Order of the British Empire at the end of 2003. They notified him by telephone, not e-mail.

A few years after he set up this hypertext network, Marc Andreessen at the University of Illinois (Go ILLINI!) took advantage of the World Wide Web and the mainly cisco routers in the network and tried to optimize their ability to deliver a stream of packets to his computer. He created the Mosaic browser, which reassembled these packets into presentable text and graphics. He would later start Netscape with Jim Clark to turn browsers into a business, but meanwhile it transformed cisco. Browsers flipped the 80-20 rule so that now only about 20% of networking was local and the rest had to go through a router to request information and packets from the Internet. Demand for routers exploded. See how simple invention can cause demand to shift rapidly? In 2003, close to 200 million NEW Ethernet ports shipped worldwide.

Not only routers were in demand, but communications links to handle the increase in traffic were as well. Modems on regular phone lines no longer cut it; traffic was too heavy. But the phone company had put in fiber-optic lines to aggregate phone lines and Internet pioneers saw those and grabbed as many as they could to handle their packets.

Intel and Microsoft, based on Doug Engelbart's blueprint, put the horizontal into the computer business. The division of labor happened almost immediately. It was cheaper to assemble chips in Malaysia than Michigan. Cisco's routers and Netscape's browsers simply rode on top of the computer industry's platform, and put the horizontal into the communications business. Without it, we'd all still be waiting for the post office to deliver information and orders for products, and we might as well have stayed in the industrial age.

• • •

Light is made up of many frequencies, bundles of photons as Einstein predicted. The Corning fiber-optic cable from 1975 was single mode, in effect, optimized as a waveguide for a tight range of frequencies. This range was wide enough, however, to carry thousands of simultaneous phone conversations. Remember, a voice call takes 64K bits per second, which is the 56K of pulse code modulation and then another 8K of other information to switch the call. Most voice conversations have gaps of silence; everybody but your mother-in-law must stop to take a breath every once in a while. So compression can be used to remove those gaps and squeeze out three-quarters of the redundant stuff, so much so that 16K per voice call is about all that is needed.

A 640-megabit-per-second fiber line between Chicago and St. Louis can (about) handle 640 million divided by 16,000, or 40,000, phone calls. AT&T developed a standard to handle phone calls on fiber optics called SONET, for Synchronous Optical Network. However, the laser signal driving those calls down the fiber line would go only about 30 miles, and about a dozen repeaters or regenerators were needed on that trip along I-55. And SONET was massively overbuilt, using rings. Two sets of fibers would send identical signals, one clockwise around the ring, and the other counterclockwise. Then, if a backhoe cut into a fiber, which would happen much too often, SONET would automatically shut off the bad line and use the good one. Remember, the phone network was engineered to go down only 2 minutes every 40 years.

Long-distance companies would then convert the optical signal back into electrical signals, read the control bits and see if any of the calls should jump off or drop from the long-distance network, to, say, connect into Springfield. Then they could add in other calls, convert the whole lot of them back into an optical signal, and blast it back onto the fiber line. This is known as OEO or optical-electrical-optical (I call it the flying monkey switches, you know, from *The Wizard of Oz*, O-E-O, O-EEEE-O). This had two disadvantages, it was costly to have all those regeneration stations, and it could only go as fast as the slowest piece of equipment. If you wanted to upgrade from 650 megabits per second to 2.5 gigabits, you had to replace every piece of electrical equipment along the path.

Solving the first problem made the second one even more glaring. In 1987, Dave Payne at University of Southampton came up with an erbium doped fiber amplifier. This was an ingenious device. An optical signal would come in and flow through a loop of fiber doped with the element erbium. Also connected to the loop was another laser that, in effect, would "pump" the old and tired signal back to maximum strength, without affecting the information encoded on the original signal. This thing worked like magic; it was an elixir that changed communications. It enabled OOO, an all-optical network.

But the amazing thing about erbium-doped amplifiers was that they would amplify a huge number of frequencies on the fiber, not just the one with the phone calls or data on it. This discovery led almost immediately to the invention of wave division multiplexing (WDM). What this meant was that more than one signal could be sent down the fiber. If different frequencies in the fiber are thought of as different colors (each color representing a different frequency of visible light), then the idea was to send different signals encoded on different colors. They wouldn't interfere while on the fiber, would be amplified together, and would just be separated at either end.

Unfortunately for David Payne, the academic, British professors rarely became entrepreneurs because the UK had a lousy history of spinning out technology from universities. It would be Ciena, a Maryland-based company, which would best exploit Payne's invention and sell some of the early erbium-doped WDM systems. Payne eventually would go the entrepreneur route, and form Southampton Photonics Inc. in June 2000, a little late to be funding optical ventures.

Oddly, WDM brought the communications industry full circle. Phone calls began on a circuit-switched network; you just took the whole voice signal and switched it from one wire to another to complete the call. Then packet switching came in, at first to prevent the vulnerability of a nuclear attack. But then packet switching took off, as the most efficient method to handle voice calls, data packets and the transport of Web pages. Packet switching is entirely electrical, a switch or router looks at the header of each packet and decides where to send it.

But now with WDM, we are back to circuits. Going from electrical to optical penalizes both cost and speed. An all-optical network means keeping the signal optical as long as possible, until the very last minute when the information is needed. Today, that "edge of the network" where optical becomes electrical is inside of AOL, or at Yahoo!'s servers, or at your local Internet service provider's facilities. But eventually, the edge will move closer and closer to users, until someday, an optical connection will be brought directly to your computer (or set top box on your TV) and it will figure out what to do with it.

But lots of things needed to be invented to enable an all-optical network. The most critical was a switch. Unlike electrical signals, it is tough (today anyway) to read information in the light without affecting it. Light must be left alone, maybe a reflection or two, but not much else.

So a somewhat dormant military technology, called

micro-electromechanical system, or MEMS, was revived to solve the problem. MEMS uses electromagnetics (thank you again Michael Faraday) to move tiny mirrors that reflect laser signals, and in effect, switch light. Early MEMS technology was 2:1, meaning you could pick one of two signals to use, but over time, MEMS was improved to take 16 fiber lines in and switch each and every one of them to one of 16 fiber lines going out. This is known as 2D or two-dimensional MEMS.

Now, a word of caution here, since we are talking about moving mirrors and electromechanics, not electrons: The switching speed of MEMS is SLOW. It is mostly used for provisioning, meaning AT&T, over a week's time, might need huge capacity to handle video from the Olympics so it would switch those lines where they are needed. Or on an hour-by-hour basis, Internet traffic might spike at news sites or MP3 downloading sites, and the appropriate circuits are set up. But these switch in seconds, not nanoseconds.

Over time, as the all-optical network is built out, 16 by 16 switches will seem ancient; 3D MEMS, which can handle a matrix of 1,000 lines input switched to 1,000 lines output, is the next invention, and then will scale up from there. If we all eventually have a color of light assigned to our homes, an array of these switches needs to be put in the backbone of the network to enable us to use it. This is a 30-year project, but you can almost guarantee it will happen. The technology is already in place.

After Sputnik, the space race heated up. On July 10, 1962 (five months after John Glenn had orbited the earth) the first international communications satellite, Telstar, was launched. Oddly, the idea for satellites came from a science fiction writer, Arthur C. Clarke. Telstar was an interesting joint venture between NASA and the Bell System, the British Post Office and the French PTT, and was used to deliver television programming. But you can imagine the Department of Defense salivating at the idea of communications satellites.

Telstar orbited the earth every 2 hours 37 minutes. Phone calls or television signals were sent up to Telstar, which relayed (receive, amplify and retransmit) them back to an earth-based receiving station. Telstar was only usable for 20 minutes each orbit to transmit signals between the U.S. and Europe. The day after its launch, on July 11, television programs were transmitted to a receiver in Pleumeur-Bodou, France, and the BBC regularly used Telstar to receive American programming, Hello Uncle Miltie.

Tracking those quickly orbiting satellites was a pain, plus the transmit window of 10–15% of the day was unacceptable. As part of his satellite dreaming, Arthur C. Clarke envisioned satellites in geosynchronous orbit. Anything orbiting the earth from exactly 22,300 miles up travels at exactly the same speed as earth's rotation, and is therefore at just the right height to always be over the same location. So now you just point your transmitter and receiver and are done. The only downside is that there is a ¼ second delay going up and another ¼ second delay on the way down. This makes phone conversations tough. But it doesn't matter for one-way usage for broadcast television, so today, all the DBS, direct broadcast satellite, systems like DirecTV and Echostar use geosynchronous satellites.

But the military wanted to use the satellites for spying and tracking. Spy satellites had the first big push. Rather than send planes at high altitude over the Soviet Union and risk having them shot down, like Gary Powers in his U2, a satellite would just innocently pass over every few hours. Cameras with super telephoto lenses could see everything. No one admits to how accurate these spy satellites are, but I once overheard an engineer boasting that he was working on a system that could count the stitches on a baseball from a hundred miles up.

But positioning systems would have far more uses. In the 1960s under a DoD contract, Dr. Richard Kirschner at Johns Hopkins Applied Physics Lab developed an early tracking system, called Transit. It was made up of seven low orbit satellites that looped over the poles, and emitted strong radio signals. The Navy wanted to use them to fix positions of ships at sea and ballistic-missile-carrying submarines. Commercial use was allowed, mainly for ship navigation, beginning in 1967 and continuing until the Navy killed it in 1996. Its problems were many: it only worked in two dimensions, it took a long time to acquire and track the signals and you had to correct for your own movement, so anything that moved fast couldn't use it.

Still, the Air Force was jealous and wanted a positioning system of its own, and funded research into System 621B (sounds like something out of Area 51). It used a new type of radio signal, based on pseudorandom noise. In 1972, using weather balloons as simulated satellites, the Air Force tested it at the White Sands Proving Ground in New Mexico. It pinpointed the location of an aircraft to within 50 feet.

Using these pseudorandom noise-based signals, and some of the first cesium-based atomic clocks that could tell time with extreme accuracy, the DoD authorized the NAVSTAR GPS system in December of 1973. It was the Cold War that helped get GPS authorized. A little known component of all GPS satellites is NUDET, or nuclear detonation sensors. These were included as part of verification for the 1963 U.S.-Soviet Limited Test Ban Treaty. It was arms control, not the need for precision weaponry that the military used as rationale for the system. Unintended consequences run amok.

Twenty-four satellites whip around the earth at an 11,000-mile altitude and broadcast the time and their precise location. When a receiver hears from four distinct sources, it calculates how long it took for each signal to arrive and therefore the distance to each satellite. From that data, it can figure out all the 'tudes: longitude, latitude and altitude. (Triangulation from three satellites would give you long and lat but not altitude, so you need to hear from four.)

The first test of NAVSTAR GPS took place in 1978, this time at the Army's Yuma, Arizona, Proving Ground—again using balloon-launched electronics to simulate satellites. The Army's involvement meant this program transcended any one division of the DoD.

In 1979, a congressional budget cut almost killed GPS,

but the DoD cut it back to 18 satellites from 24 and pushed out delivery dates. The Space Shuttle had been chosen as the cost-effective vehicle to launch GPS. On January 28, 1986, at Cape Canaveral, the next generation or so-called Block II satellites that are the backbone of GPS today were about to be launched. It was a beautiful sunny morning.

But tragedy struck. The Space Shuttle *Challenger* exploded 73 seconds after takeoff, killing all seven astronauts, including schoolteacher Christa McAuliffe. The accident delayed any new shuttle launches for two years. The DoD scrambled to use Delta II rockets to launch GPS satellites, and by 1988 and 1989, they were back on a new schedule for deployment.

It is unclear if the military ever intended that its new toy would be used by, say, an absentminded corporate exec to figure out where he parked his car. But after Korean Air flight 007 strayed over Soviet airspace and was shot down, President Reagan encouraged GPS usage for civil aviation. It was early enough in the project that hooks for commercial usage could be added.

From that time, GPS specifications added commercial applications to strict military usage. There is CA (Coarse Acquisition), which most tests show works down to an accuracy of about 5–15 meters and PPS (Precise Positioning System), which is accurate to probably a foot. The PPS mode sends a highly accurate but encrypted signal, so only authorized users with the right "keys" can use it. There is also an SA or Selective Availability mode; in other words, DoD can degrade accuracy to 50–100 meters from 5–15 for Coarse Acquisition users, in times of war. Or when they feel like it. Until May of 2000, SA mode was usually kept on.

The Gulf War and Operation Desert Storm in 1990 and early 1991 proved the value of GPS, even though only 16 satellites were up and working. Helicopters, fighter jets, and tanks all used GPS receivers. CNN showed video of receivers duct-taped to the side of half-tracks. The DoD put in an emergency order for 10,000 units from commercial as well as military suppliers. Trimble Navigation had a huge backlog of orders, some from the military, but many from soldiers or from families of soldiers for personal use so they wouldn't get lost in the desert; a very strange twist of personal purchasing and use of commercialized military technology intended solely for military use. So many commercial receivers were in use during Desert Storm that the DoD had to turn off Selective Availability so soldiers could get accurate readings.

By the time of the 2001 war in Afghanistan and the war in Iraq in 2003, GPS was built into guided missiles and bombs. Troops on the ground could acquire coordinates of a target, and relay them up to B-52 bombers who would program precision-guided weapons to explode exactly where they wanted. Given statistics that 90–99% of bombs dropped during World War II and the Vietnam War missed their intended targets, GPS has changed the face of warfare.

But it has also changed the commercial world. Honeywell developed a navigation system based on GPS so that big commercial Boeing aircraft could land themselves. Hardly a ship or yacht goes out to sea without a GPS system. Some hikers live by them. GPS-enabled watches can now track a hiker's path and a shareware program can download satellite images from the trail and reconstruct a photo tour of the outing.

Single chip GPS receivers are being designed into cell phones, so that users can be told exactly where the nearest Starbucks is. Precision guided marketing. The commercial acceptance of GPS is quite telling, and not only the DoD, but also the DoT. The Department of Transportation now oversees its implementation.

The need for precision weapons would both directly and indirectly launch the digital revolution: transistors in 1948, lasers and integrated circuits in 1958, packet switching in 1964 and microprocessors in 1970, and that was just the easy stuff.

Computers were invented to help win World War II. John von Neumann and the Moore School at the University of Pennsylvania designed the ENIAC digital computer, the birth mother of the U.S. computer industry, to speed up calculations for artillery firing tables for Navy guns. At the same time, Alan Turing and the British at Bletchley Park designed the Colossus computer to decipher Enigma codes. A host of electronic devices at Los Alamos helped speed up difficult calculations controlling the reaction of uranium–235 for the atomic bomb. The Space Race gave us much of the rest. Of course, it was commercial use that drove progress down the cost curves, but often, like it or not, conflict sparked invention. Ask Wilkinson and Watt.

So now, most of the piece parts are in place. Logic and Memory in microprocessors. Packet switching to break up pipes and routers to move these packets around. Fiber optics with information carried on photons. And radios that can carry packets of data instead of just voice calls or AM radios' traffic and weather together every 10 minutes. From these pieces, there are so many other stories to tell, of wireless data, Open Source software and the many other forms of intellectual property that can be efficiently and productively put together.

PART 5:

MODERN CAPITAL MARKETS

Trade in technology today is not much different than it was in British textiles. As long as microprocessors and digital electronics get cheaper, it is a win-win for the producers and the consumers, as the cheaper functionality replaces something else. But as we saw with the British, how you get paid can help or hurt the creation of wealth and improvement in living standards. Today, money sloshes around, but back in the 18th century money was a fairly local instrument. Precious metals such as gold and silver were the de facto currency for trade. But monarchies and their governments created their own currency, backed by gold. First as a convenience since a titan of industry would need wheelbarrows filled with gold to do his business, and then as a tool to control the economy. Today, the world doesn't think about gold much, except around birthdays and anniversaries.

The British pound's convertibility to gold, predictably, was suspended during World War I. It took another seven years to get back on the gold standard. A guy named Churchill put England back on the gold in 1925, at the prewar exchange rate. Heck, it was still Isaac Newton's rate. Big mistake. It overpriced the pound, which got dumped, and gold quickly flowed out of the country. Banks had restricted money supply and England went into a nasty recession, a loud advertisement for floating exchange rates.

By 1928, the rest of the world went back on the gold standard, but not for long. Following the 1929 stock market crash, 1930 saw the introduction of the protectionist Corn Law–esque "Smoot-Hawley" tariffs (some say the market predicted the tariffs, which were debated in 1929). Trade dried up as most other countries put up protective trade tariffs and increased taxes to make up for lost duties. A recession began in the U.S. and elsewhere, and the Federal Reserve, formed in 1912 to act as the central bank for U.S. currency, in effect mimicked the Bank of England. The Fed forgot, or didn't know, that in a fractional reserve banking system, it was the lender of last resort, a concept that Brit Walter Bagehot had brought up years before.

In February 1930, it did cut the rate it would lend money from 6 to 4%. It also did expand money supply to a small extent. But the Fed chairman insisted that the situation would work itself out. That year saw the first wave of bank failures and each time a bank failed and deposits lost, the money supply was shrunk by that amount. By 1932, some 40% of all banks—roughly 10,000 of them, and $2 billion in deposits—disappeared. The money supply dropped by over 30%. So did the gross national product, and just like that, 13 million people were out of work. If the Fed had just provided money as lender of last resort, most of the carnage might have been avoided. But no, the U.S. was back on the gold standard and not allowed to increase the money supply, lest gold leave the country.

Herbert Hoover vacuumed up whatever capital was left by increasing the top tax rate from 25 to 60%. In response, Franklin Roosevelt campaigned with "I pledge you, I pledge myself, to a new deal for the American people."

In March 1933, just after FDR's inauguration, unemployment hit 25%. After yet another bank run Roosevelt declared an eight-day banking holiday after which confidence in banks returned and deposits flowed back in. Later in 1933, the U.S. dropped the gold standard, following England, which dropped it in 1931. Unshackled, money supply could now increase and replenish banks. After a yearlong recession in 1938, New Deal spending kick-started the economy. A world war kept it going.

Meanwhile, the Germans were stuck paying World War I reparations. In 1921, the victors presented a bill for 132 billion gold marks. With a crippled industrial base and dependency on imports for raw materials, the Germans were forced to print money to pay back debts. The mark was devalued by a factor of 481 billion. By November of 1923, the exchange rate was 4.2 trillion marks to the dollar. While the resulting hyperinflation left Germany without debts (and debtors learned a lesson to carry debt in dollars!), it wiped out the country's savings and put the German economy into a severe depression, leaving the country susceptible to new political regimes.

On July 1, 1944, just 24 days after D-Day and well before World War II ended, a dollar standard was put in place, as a result of the Bretton Woods agreement. It was considered a gold standard, but even economist John Maynard Keynes called it "the exact opposite of the gold standard." The U.S. dollar was pegged to gold at $35 per ounce, and became the only currency allowed to convert into gold. The rest of the world's currencies were pegged to the dollar at a "sort of" fixed rate. "Sort of" because the rate was fixed and stable until there was some trade problem, and then the currency would be quickly revalued to a new official/stable exchange rate. Stable and floating—quite the paradox.

Banking continued as England or France would hold dollars, and lend against them as reserves.

The problem was the dollar itself. The Vietnam War and LBJ's Great Society welfare program helped create inflation in the U.S. Like the bank runs in England from having too much gold, and with inflation devaluing money, so too the U.S. had a run on the bank. In 1963, gold reserves in the U.S. were less than foreign liabilities—by 1971, they were barely 20% of the amount needed to cover liabilities. The run started as dollars were dumped, and despite efforts to kill inflation and shore up the dollar, Nixon halted convertibility on August 15, 1971. Finally, elasticity could run free of the restraints of gold. Mark this, then, as the birthday of the modern world.

With flexible exchange rates and relatively free trade in post–World War II (and to be fair, post the Bretton Woods gold standard), low-margin tasks and low-paying jobs moved out of the U.S. Displaced workers and union rhetoric screamed for something to be done about "lost jobs" but it probably was the best thing to happen to the U.S. since it allowed for high wages in the U.S. for high-margin tasks. The stock market rewards high-margin companies with high values, lowering their cost of capital. They can sell fewer and fewer shares to raise money instead of borrowing money from banks. High wages are taxed, to pay for social services, not the least of which is a military to protect the U.S. AND its trading partners.

But a new twist was added to this system. Since the birth of the personal computer and the horizontalization of the industry, companies could focus on thin slices of intellectual property, which could have very, very high margins. Software, microprocessor architectures, semiconductors, network architectures, optical components, cell phone components, and databases are all pieces of intellectual property that could be licensed to others to build end products.

In fact, the U.S. is a huge exporter of these pieces of intellectual property, although these exports are hard to measure. Often, an entire architecture of a chip, valued at a billion dollars, can be e-mailed to a factory in Taiwan, without a cash register ringing or a Commerce Department employee around to measure the export. That chip and other intellectual property are then combined, using low-cost labor with other low-margin components, like a power supply and some plastic, and turned into a laptop or DVD player. Oddly, this "margin surplus" run by the U.S. is the way to run an economy with declining price products. Gold won't help. Instead, it requires a stock market to balance out world trade. Fortunately, we've got one of those!

The Business of Wall Street

Calvin Coolidge said that the business of America is business. What the hell did he know? He never worked on Wall Street. The stock market is simply about access to capital, and Wall Street provides world businesses access to the stock market, for a not so modest fee. Careers on Wall Street are amazingly lucrative.

But how did we get from Elizabeth I's Royal Exchange and South Sea bubbles to today's billion-share-a-day trading, tech IPOs and derivatives to hedge the weather?

Government control of money is often a mistake, as politics rarely gets what is best long term. Elizabeth I was smart about this. It would be difficult for anyone to properly issue just the right amount of money to grow the economy, and the best structure would be a bank operating at arm's length that could handle this function. The British smartly set up the Bank of England, which sometimes competed with other banks, but would usually complement them by lending what was needed.

But banks are funny. As we saw, they are bankrupt, almost by definition. Worse, they are risk averse.

The stock markets, or exchanges set up to trade in shares of companies, love risk or at least appreciate it. In it for the upside, they don't worry about getting loans paid back. But it took awhile to get going. For almost two centuries after the start of the Royal Exchange in 1566, not much trading was done there. A few rowdy brokers got kicked out in 1760 and set up the London Stock Exchange, just in time to help finance the Industrial Revolution, at least in a small way. And ever since the New York Stock Exchange got its start in 1792, it has been a much bigger contributor to American innovation. More so NASDAQ after the U.S. markets partially deregulated in the 1970s.

Money makes the world go round, but the stock market decides who gets the money. Is that all the stock market is good for?

Say you own 10% of a socks factory, but need to raise money for your cow farm. So the stock market provides liquidity, cash for your share of the business. In doing so, it is also a great, nongovernment mechanism for setting the price of businesses and allocating capital to them. The market will pay more for shares of a business that it expects to have a bright future and as a result the company can sell a smaller percentage and raise the same amount of money. In effect, its cost of capital goes down. On the flip side, the market can starve bad businesses of capital to stop them from throwing good money at dead-end operations.

But funding ideas and intellectual property couldn't be more different than funding factories. Building factories and stocking them with machines requires huge outlays. Buying raw materials and paying for them well in advance of getting paid for finished goods requires big chunks of working capital. Building inventory and lining up and paying for transportation for the raw materials coming in and the finished goods going out adds tremendously to costs. I get a headache even thinking about the logistics nightmare of a factory owner in 1820. Add to that the cost of insurance, even if it were available, and costly waste and it is a wonder that factories made any money at all. But they did, because before the Industrial Revolution, making products by hand was an even more expensive endeavor. Cottage industries produced a low quantity of low-quality goods.

Today, industry still exists; it will never go away. But it is well capitalized. The quest for profits comes less from squeezing costs out of the manufacturing process and more from expanding sales channels, or creating new designs, or leveraging other partners or lowering customer service costs. These require computer systems and software tailored for specific tasks, rather than a new factory.

The role of the capital markets today is to provide the funding for those electronic systems and the capital required to pay people, rather than to fund inventory. A lease on some building space, a few computers, fast Internet access and a manager or two and you are off and running, producing code to save the aluminum and chemical industry from extinction.

The biggest problem facing any new business, be it steam engines or static memory, iron foundries or semiconductor fabs, is finding capital to fund the business. Banks won't lend money to businesses they don't understand. That's because their method is to study the past financial history of a business to predict the future cash flow and the likelihood they will be paid back. New business? Forget it. Come back when the profits roll in.

Back in 1769, Matthew Boulton was attracted to James Watt and his steam engine and provided him with risk capital because he understood early how the steam engine could change the manufacturing business. In exchange, he got two-thirds of the business, which reaped him 25 years of dividends.

A stock is nothing more than the current value of all of those dividends. Boulton and Watt could have gone public, and the value of their stock would have been the sum of the dividend payments, adjusted for time and risk and competition. Boulton could have cashed out in year three, and many others could have owned his piece of the steam engine franchise. In fact, Watt could have cut out Boulton altogether, and just sold a piece of his business to the stock market, and used that capital to fund the business. But there wasn't a vibrant stock market back then. There is now, and it funds a Watt a week.

Today, it is venture capitalists who provide the very early stage risk capital, but very few have the patience to hold on to their share of a business for 25 years to reap the dividends. Fortunately, there are plenty of other investors who will hold on to businesses they believe in for decades. Pension funds, university endowments, and foundations are all institutions that are willing to be long-term investors. They will own them for 20, 30, maybe 100 years. Mutual funds that try to grow their investors' money faster than the overall market will hold shares for 2–5 years. Hedge funds, on the other hand, since they look for quick changes in value based on short-term information like inflation data or pricing trends, might own stocks for 2–5 minutes.

The stock market is a great mechanism to:

1. Provide expansion capital for businesses
2. Agree on a price for a business
3. Transfer shares from owners to others who may have a completely different risk profile or time horizon

Wall Street, and trading through people and specialists at stock exchanges, has always been great at getting stock into the right long-term hands, for a fee. That was fine when there were a few hundred interesting companies in 10 unique industries, and trading was around the edges. But in 2002 there were thousands of interesting companies, in probably a thousand unique industries (I can think of a hundred different segments of the software market alone).

From its debut as a hot IPO, the Bank of the United States was capitalized with $8 million. The year before, in 1790, the new government of the United States of America sold $80 million in bonds to pay for the Revolutionary War and General Washington's bar tab. Trading scrip on the muddy streets was no way to go through life, so the first organized stock exchange in New York was formed May 17, 1792, when 24 brokers and merchants met under a buttonwood tree that has since been replaced by a building at 68 Wall Street.

Two centuries ago, standing on a soapbox was considered high tech. A stock exchange could not be much larger than someone's voice could carry.

These guys were hungry for action and someone had to move those bonds and scrip around. This group became the New York Stock and Exchange Board, and all sorts of bonds and other bank stocks began to change hands there. Traders made money via spreads. They would buy a stock at par value, say $100, and sell it at a $1–2 premium.

After the Bank of the United States IPO mania, things settled down and got pretty dull. Stocks traded by appointment, literally. Twice a day, someone would slowly rattle off the names of stocks, and if anyone wanted to trade, they would do so after their names were called. In effect, the New York Stock Exchange, nicknamed the Big Board, created a cartel. By controlling the trading in self-defined "listed" shares, others couldn't come in and low-ball their spreads and commissions. This monopoly was extended to member firms only, and required a license to trade at the exchange in listed shares with fixed commissions. The downside was a limit on liquidity, but the "members" didn't care. That wasn't their problem.

There was leakage from the New York Stock Exchange. Further down Wall Street, traders would occasionally sneak out and trade "off-exchange." These fixed commissions stuck until 1975. The monopoly was supposed to have ended in 1975 as well, though in practice it still exists today.

Congress closed the Bank of the United States in 1811. Seems as if a few "friends" of congressmen wanted in on the loan business. Not surprisingly, a slew of state banks took up the slack, and many saw their stock trade at the NYSE. Another Bank of the United States was charted in 1817, but it appeared to be terribly mismanaged. Andrew Jackson took office in 1829 and was convinced the Bank of the United States was corrupt and unconstitutional (banks aren't mentioned in the Constitution). Jackson vetoed the bill that would extend the bank's charter, set to expire in March 1836, and withdrew U.S. Treasury money from the bank. For his efforts, the Senate censured him. But since he ended up with his face on the $20 bill, he got his day in the long run. State banks and local banks looked at the absence of a central bank as a license to take over and lent money willy-nilly, to all takers. Speculation ran rampant in America in 1836. Out on the streets once again, people were trading stocks, bonds and any instrument they could.

In 1836, the Exchange took action and forbade its members to trade out on the street, since that clearly threatened its fixed commissions. Credit tightened in 1837, and a Financial Panic took hold, ending the speculation. Hundreds of banks failed, trade businesses in New York shuttered, even the U.S. government had difficulty paying interest on its debt. It took several years for commerce and the economy to get back on its feet again. This was not the first or last time that speculation followed by panic created huge ups and downs for the capital markets, but each time the stock market survived and went about its business of providing capital to growing businesses.

By providing liquidity for government debt, banks and soon industrial businesses, the stock market was providing a great service to the U.S. economy. Money was no longer "stuck" in fixed asset investments. By selling stock to someone else, say a bank that wanted to own a portfolio of factories, capital could be freed for investments in new, riskier businesses, with potentially higher returns. This is a stock market at its best, where capital moves around according to the risk profile of each participant.

In the first half of the 19th century, transportation was the rage: Turnpikes were funded in the early 1800s and canals were funded between 1820 and 1850. But then in 1844, investors saw the introduction of the Morse telegraph and an era of faster and cheaper communications began. Wall Street immediately saw the benefits. The telegraph gave investors outside of New York access to more up-to-date pricing information. While the area under the buttonwood tree didn't get bigger, the telegraph funneled more cash into the exchange. Technology constantly increased the speed of information and speed meant more profitable trades. Telegrams were invented to relay messages. The most amazing stat I found suggests that trades in stocks and bonds accounted for over half of the telegraph usage in those days.

Ezra Cornell wired up the East Coast with telegraph lines and consolidated them into Western Union in 1855. Markets were more interlinked than ever. In 1857, Prussian wheat was dumped on the world markets, hurting the U.S. railroad's wheat transportation business. This slowed the expansion of the railroads, which hurt western land speculators, such as the Ohio Life and Trust Company in Cincinnati, which promptly collapsed. News spread via telegraph, causing a run on U.S. banks, which didn't have enough gold to give depositors because the gold-carrying ship *Central America* sank in a hurricane. All those falling dominos led to the Panic of 1857. Ouch.

In 1861, Western Union turned on the first transcontinental telegraph connection, finally closing the Pony Express. This helped move information even faster to the markets that needed it. Telegraph stocks were the play in the 1860s. Hot, hot, hot.

Trading volume grew, but capital committed to investments wouldn't see the same increase until 1866, when a Brit named John Pender came up with tough enough telegraph cables that could withstand the sea, licensed from the Gutta Percha Company and laid by I. K. Brunel's massive steamship, the *Great Eastern*. Pender's company eventually became Cable & Wireless Ltd.

For the first time, people in London could talk, actually key, someone in New York almost instantaneously. This beat transmitting prices by ship and weeks of delay. Plenty of other companies raised money to fund their transatlantic telegraph cables, on Wall Street, as well as in London and Paris. International telegraph companies charged around $5 per word and could transmit 15–17 words per minute. A company could make a couple of million dollars a year, and investors were hungry to provide risk capital to be a part of it. Within five years, prices dropped with competition, and many of these companies went bust or were bought up on the cheap. Global Crossing would lay fiber-optic cable under every ocean 130 years later, and go bust when prices collapsed from competition. History repeats.

But since the Brits the first time around were sitting on piles of pounds as the Industrial Revolution engine revved (and the flow of gold inundated banks with huge reserves), these undersea telegraphs opened up American markets in a bigger way to British investors. Just in time too, as the railroads required lots of capital to fund the extension of their tracks. The 1860s and 1870s saw an enormous amount of money wired to build railroads. The British seemed to prefer debt to stocks, as they wanted to be paid back their principle. Their South Sea Bubble memories and aversion to risk would cost them.

The year 1867 saw the introduction of stock tickers, a modified telegraph device. As volume increased, the twice-a-day roll call changed to a specialist system in 1871. A specialist would be given a monopoly in the trading of a given stock, in exchange for promising to maintain an "orderly market." What this meant was that the specialist had to provide liquidity—buy when there were too many sellers and sell when there were too many buyers—to keep the stock price from experiencing wild swings. If you wanted to trade Amalgamated Mogul, you had to go to the specialist's trading station and make a bid. The specialist didn't charge a commission. But he did charge a hidden fee disguised as a spread. He just bought at one price and sold at a higher one.

That sounds fair, but providing the liquidity still entailed risk. So specialists would trade for their own account, under the premise of providing liquidity. This is perfectly legal under the specialist system. Think about it, this institutionalized an asymmetry of information. By guaranteeing to provide a very necessary liquidity (which technology will eventually step in to provide, but we're still a hundred-plus years ahead of ourselves), a specialist knew the real price of a stock and outsiders didn't. He got the flow and he got the trading profits. This centralized liquidity was the right thing to do in 1871, yet despite a millionfold increase in the speed of price information since then, it still exists. Today, liquidity is formed out at the edges, yet the New York Stock Exchange clings to this profitable trading structure.

Prices on the latest railroad stocks traveled over telegraphs and were punched out on these new stock tickers, and more money flowed in to fund the railroad buildout. It got so out of hand and speculation was so rampant that even lousy railroad ventures were being funded. Jay Cooke and Company, a respected banking firm in Philadelphia that helped fund the Civil War, was almost single-handedly throwing good money after bad into the North Pacific Railroad. Deflation hit the U.S., and in 1873, when the North Pacific failed so did Jay Cooke, and the Panic of 1873 whipped through the country. Wall Street reemerged yet again from this financial crisis, but this time insisted on even more information, not just stock prices, but news from companies, so it could figure out what to fund, and what not to fund.

It got what it was looking for with the invention of the telephone. The year 1878 saw telephones on the floor of the exchange, when a specialist picked up a phone and said "Buy-bid-'em-up-sell."

In the 19th century, the U.S. population was growing like a weed, filling in the wide-open spaces out West, and following the British industrialization, albeit with a 50-year lag. All the stock market had to do was provide capital. Whatever the Street could skim off was fine. The spreads the Americans suffered were more than worth the added cost of the capital, because the returns on industrializing were so huge. Sure, there were a few panics now and again, but just the existence of capital made America grow. I wish it were that easy today. The stock market today is even more important to economic growth, and the skim a serious detriment.

Unlike a bond, which represents a debt of a company or government, and will (please, please) be paid back, stocks represent an ownership stake in a company. Valuing a company is an art not a science, but in a sense, stocks simply represent the current value of future dividend payments, with the big honker assumption that dividends are paid out of profits.

Without computers, who the heck could even figure out dividend ratios let alone price to earnings multiples? Hence a stock would trade as a percentage of its par value. The yield, or a company's future dividend payment, was subject to debate, so stocks were always more speculative than bonds, but they traded like bonds. A trader would buy a stock on the exchange at 83% of $10 par value, but sell it to a broker at 85% of par value who would then sell it to a wealthy customer on Fifth Avenue at 88% of par value. Not a bad day at the office. A 2% spread for the trader and a 3% spread for the broker.

Eventually, the wealthy customer would figure out how badly he had been ripped off, and install a stock ticker in his parlor overlooking Central Park. Magically, the spreads, or the markup between buy and sell, would shrink. But lo and behold, every time the spread would shrink, the volume of trading would increase. It's as elastic as cotton and transistors!

After the ticker and the telephone, not a lot changed on Wall Street. Volume edged up. The railroads were built, financed with Wall Street capital, much of it brought in from Brits chasing higher yields. A panic here or there, but trading worked relatively smoothly and capitalism prospered with this steady source of capital. In 1886, the NYSE saw its first million-share day, an almost 1,000-fold increase in volume from the 1,534 shares that traded in June 1837. Technology worked in favor of Wall Street: The more people connected to the exchange, with fixed commissions, the more profits to be had. The only hassle was that when you traded a million shares you had to get the certificates from the sellers and deliver them to the buyers in exchange for payment. This would eventually haunt Wall Street and become the biggest driver for computerization of the stock market.

In 1884, Dow Jones began publishing end-of-the-day stock quotes in a newsletter that eventually turned into the *Wall Street Journal*. The *WSJ* brought quotes and news to the chunk of the population who couldn't afford their own stock ticker. Stale, day-old quotes worked for many people, and Wall Street was happy to have people "think" these were real prices.

Most of the use of technology by Wall Street was to handle its internal workings. With fixed commissions, technology wasn't needed as much to drive revenue as to keep costs down. Pneumatic tubes were installed at the exchange in 1918, mainly to send completed trade tickets from the traders to the back office clearing humps. In April 1920, the Stock Clearing Corporation was created to handle the increased paperwork. This was mainly a large room with piles of certificates passed among workers.

So now, on the brink of the booming 1920s, Wall Street had what it needed to fund America's post–World War I growth. Banks provided mortgages; ships could handle trade; trucks coming out of the factories could move goods around; cars became affordable for the nouveau riche; and soon proliferating radio stations blasted culture high and low into living rooms. Again, the stock market served a valuable function by providing risk capital to these new ventures. And investors were happy to chase the potential returns. The titans of Wall Street at the time, such as J. P. Morgan, were well compensated for their ability to provide money for these companies. Everybody benefited. And I don't mean just the elite who actually invested, but also workers, as investments in railroads and steel companies created higher-paying jobs. Speculation and panics were part of the process. J. P. Morgan held off panics in 1893 and 1907 by putting in big buy orders to the exchange when there were too many sellers and specialists didn't have deep enough pockets.

After World War I, Americans were enticed to invest on Wall Street with not just the high returns these new ventures offered, but with leverage—scale like the big boys had. With 5% down and 95% borrowed you could be as rich as J. P. Morgan before age 30. This speculative leverage enhanced what would already have been an amazing bull market. In September 1929, Yale Economics Professor Irving Fisher, the market strategist of his day, said he thought stocks would "remain permanently high" (oops). It all ended on October 24, 1929, with the Great Crash. Since that day, so the legend goes, no buildings on Wall Street have been built with windows that open.

The next day, 12 million shares traded and another 16 million a week later. Wall Street may have weathered the storm. Unfortunately, the Smoot-Hawley tariffs on imports and a Hoover tax hike sunk the boat in 1930.

The Crash of 1929 and subsequent Great Depression gave Franklin Delano Roosevelt and friends a free ride to write a lot of rules, most of which we are sadly still stuck with today. The Securities Act of 1933 provided for full disclosure by companies raising capital. The Banking Act of 1933, cosponsored by Virginia Senator Carter Glass and Alabama Congressman Henry Steagall, erected a wall between banking and the securities business that would last until the end of the century. The Securities Act of 1934 created the Securities and Exchange Commission, which like a hand from the grave of worse times, still controls how Wall Street operates, how initial public offering documents are written and marketed, and how information is disseminated.

As for stocks, commissions were fixed, lunches were long and afternoons were short. Volume languished for almost 20 years. October 10, 1953, saw 900,000 shares traded. That was the last day that fewer than 1 million shares traded at the NYSE.

Computers and faster communications would change all that and create a Modern Stock Market. But at the same time, it would make obsolete many old ways of doing business on Wall Street.

Let's jump ahead for a second. What does a modern stock market, I mean one with computers, look like? Today it means automated trading via matching services and over half of NASDAQ trades done computer to computer. Something has clearly changed, but not enough to cancel the Buttonwood Agreement. The structure of Wall Street directly affects the costs of raising capital. Get its structure wrong, and we will soon be pricing securities in another currency.

Who determines the structure? Good question. At its most basic level, Wall Street facilitates companies turning in colorful pieces of paper, stock certificates, in exchange for real money to fund growth. The fact that those certificates trade, and increase or decrease in value, is the Grand Illusion and the secret of success of Wall Street. The structure of the Street is ever changing as its participants figure out the most efficient, and real-time ways of dealing with each other. The government steps in with rules and regulations when Wall Street's greed blatantly hurts individuals; when the playing field tilts too far. But increasingly, participants, the NYSE for example, use regulations to freeze the way business is done. This is the 30-second rule that I'll explain later, and it inhibits new efficiencies.

The necessity of raising risk capital for new ventures will be more important over time, not less. Unfortunately, the old structure of Wall Street imposed greater and greater costs to raising capital, hidden taxes that slow entrepreneurial growth. Left standing, capital will go elsewhere.

The sweeping and rhythmic dance pairing technology and capital markets has been going on for 152 years. Wall Street has a love affair with technology, to obtain valuable information faster than others, to handle crippling volumes of transactions, and then to invent profitable products. In fact, Wall Street and technology are as inseparable as, uh, Bonnie and Clyde.

One of the earliest technology providers to capital markets was Reuters, which famously got its start in 1849 by transmitting stock prices by carrier pigeon between Aachen and Brussels, thereby outracing messengers on horseback and word of mouth. Investors paying Reuters for its service could profit from buying or selling mispriced securities, mispriced because the sucker on the other end of the trade didn't have as up-to-date information. There has always been a thirst for speed. Ben Franklin opined that time is money. This is also every trader's credo. From its inception, the New York Stock Exchange operated on the principle that speed dissipates as you move further away from the trading floor.

But, Reuters's "pigeon advantage" didn't last long. There were no regulations stopping others from running telegraph lines between Aachen and Brussels. In effect, they would eat the pigeon for lunch. Reuters then ran telegraph lines to bring the greed of speed to its customers. Investors would eventually buy and overpay for shares of telegraph companies. This provided the needed risk capital for growth. The dance between capital markets and technology continues today.

But while markets are all for taking risks, almost everything else is dead set against it.

I guess I just don't get it. My health insurance company, which pays my medical bills, would probably prefer I die if I get too sick and too expensive to keep alive. My life insurance company, which pays a lump sum to my heirs when I die, would probably prefer I stay alive forever. Isn't it all backwards? Shouldn't my life insurance company bend over backwards to keep me alive? And shouldn't my health insurance company be incented by having to pay a huge penalty/payout if I die? It all depends on who pays for the risk.

Having Wall Street around to fund technology accelerated its usage. But as we all know today, technology is not without its risks. Risk of the technology not working, sure, but risk from obsolescence is an even bigger factor. Stock markets are usually good at assessing risk (but not always, I lost track how many bubbles and panics I've already mentioned!) and adjusting returns accordingly.

The other side of the stock market's risk profiling is the insurance business, which does just the opposite. It assesses the risk, and puts aside what it needs for a rainy day. It doesn't provide risk capital. It takes it away. Risk, at least innovation risk, is what drives progress. If insurance is a drag on innovation, we better figure out why.

• • •

Trade is as old as Adam and Eve, the whole rib thing and all. But as far as I can tell, Adam didn't hedge his bets by offering a fibula as an insurance premium to get out of it in case the deal went bad.

Insurance is almost as old. In Babylonian times, caravans were constantly plundered, camels strewn asunder. Lenders to these caravans would charge a premium, in addition to the probably usurious interest rates (some say 20–30%), and cancel the loan if the caravan was robbed. Of course, you didn't have to pay back the premium, but the collateral for the loan was the lives of you and your family; you'd all end up as slaves. The Code of Hammurabi put property laws in print, but then again, there seemed to be more codes dealing with the "night trade" than with trading in food or other property. You can imagine that most traders on these caravans paid their premiums. The lenders offset their risks by lending to many caravans.

Ancient Greek and Phoenician merchants would borrow money to pay for crew and cargo on dangerous excursions on the high seas. Again, in Solon's Greek laws, like Hammurabi's, the penalty for property loss was slavery. A contract or "respondentia" was set up, with premiums paid above and beyond the interest. These civilizations continued the tradition of premium payments and loan cancellations upon loss of ship and cargo, adding the risk of weather to the risk of thieves. These respondentia contracts were often sold to third parties, an early form of securitization.

It makes sense that the existence of insurance actually increased trade, which is what made these civilizations rather than chaosizations. More merchants could take risks, and sleep better at night, without gambling away their lives. For this, society benefited.

But between then and now, something has gone terribly

wrong with insurance. It seems as if insurance now touches everything. It is hard to turn around without bumping into another form of insurance, often mandatory. Instead of promoting risk taking, insurance has made society risk averse by taking risk capital OUT of circulation. Let's see how.

Those early lenders collecting premiums didn't know about the probability of loss. Probability hadn't yet been invented. Sure, they had a gut feel for the likelihood of a caravan being robbed or a squall hitting a ship of Grecian urns, but they were just making it up.

Our hero Blaise Pascal did more than invent the calculator for his tax collecting father, lighting a slow fuse on the computer revolution. He also proved that vacuums exist and that pressure could be measured with a tube of inverted mercury—two phenomena that James Watt needed to get his steam engine working. But Pascal's rapidly firing mathematical mind would go to other, more near-term pursuits. He was a vicious gambler. He had to know when to hold 'em and when to fold them, as well as calculate the odds on any given roll of the dice. Along with Pierre de Fermat, whose Last Algorithm puzzled math-heads, Pascal struggled on one particular gambling puzzle: how two dice rollers would split the stakes or bets if they left before finishing the game. In 1654, Fermat and Pascal sent letters back and forth and solved this puzzle inventing, in the process, the field of probability, on which risk management is based.

Of course, it is often impossible to distinguish insurance from gambling. Pascal wrote:

> the first thing which we must consider is that the money the players have put into the game no longer belongs to them . . . but they have received in return to expect that which

luck will bring them, according to the rules upon which they
agreed at the outset.

Now read it again; that sounds exactly like the auto insurance I am forced to buy, or my health plan coverage. Insurance is forced gambling with someone else besides you and
me playing the house, setting the rules and collecting the pot.

But that is not my beef with insurance, heck, you and I
can just turn around and become insurers, or invest in insurance companies and become the casino rather than the gambler. Instead, my beef is what insurance companies do with
the money. In effect, they stick it under a mattress, which
leaves less for innovators, who take business risk, to try new
things.

Other famous scientists played around with probability theory. Huygens, the Dutch philosopher who conceptualized the
gas-exploding piston engine, wrote a book on probability in
1657. Leibniz, who added multiply and divide to the Pascal
calculator, also played around with probability, where it inter-
sected with the law (does it?).

But the main use for probability was still for insurance,
and specifically marine insurance, which grew in importance
into the 18th century. British merchants were plying the Triangle Trade, borrowing heavily for their ships and crew and
working capital. The East India Company's dangerous trips
around the Horn to India and China needed insurance-
backed funding.

After the Restoration in 1660, commerce accelerated in
London. The official place to do business was the Royal
Exchange, but this stuffy, inflexible institution was no match
for side deals done out on the "street." Well, not literally the
street. There were a few warmer places to meet in London

to do business, the pubs and the coffee houses. For some reason, more business was done by the caffeinated than the quaffed.

One such coffee house, Quaker Edward Lloyd's, became the center for marine insurance. It was first mentioned in the press, located on Tower Street, in 1688, the same year William and Mary began their reign, and it moved to Lombard Street in 1791. Merchants, through an "insurance office" or broker, would seek out wealthy individuals to sign a policy, taking on a percentage of the risk of a ship in exchange for a similar percentage of the premiums paid. They would sign at the bottom of the policy, under the wording and under the previous signer's name, hence the name underwriters. Many believe this signing under dates back to 14th century Genoan merchants. But who really cares? It's a stupid name that stuck.

Business at Lloyd's became huge. Edward Lloyd himself made money selling coffee and reliable shipping news. Those who frequented his coffee house prospered as well. The British Navy didn't rule the seas quite yet, but was strong enough to protect shipping. Someone must have been in the back office with a Pascal/Leibniz calculator figuring out probabilities and what size premiums to collect to make up for risk.

It was all very informal until 1720 when the South Sea Company's stock collapsed with a spectacular thud, taking thousands of dumb investors with it. Like all good scandals, the South Sea Bubble spurred the government to fix an unbroken system. (It was the South Sea Company that was broken, not the system.) Most investors saw their holdings rolled into consolidations, or consols. And Parliament passed the Bubble Act, which forbade any business except the Royal Exchange Assurance Company and the London Assurance Companies from selling marine insurance.

The folks at Lloyd's obviously had great lobbyists, because there was a huge loophole. Individuals could become underwriters by taking on risk, according to the Bubble Act:

> [E]ach for his own part not for one another and, by long standing custom, the whole of their private estates was pledged as security to meet a claim.

OK, so Oliver Twist wasn't going to become an insurance underwriter, but wealthy individuals backed by their real estate and other holdings could write and trade insurance. They could use illiquid holdings to get very liquid premiums. The coffee house was spared, and the gang at Lloyd's continued to do the sea lion's share of insurance underwriting. To show how long most government regulations outlive their usefulness, the so-called Bubble Act was not repealed until 1824.

But insurance and gambling were still tied together. Lloyd's and other coffee houses had royal death pools and bookmakers for just about anything. In 1769, 79 "members" representing most of the marine insurance business moved around the corner into New Lloyd's Coffee Shop (these Brits love the concept of new, New York, New Jersey, New London, New and Improved).

In March 1774, Lloyd's, the generic name for London-based marine insurance, rented rooms back at the Royal Exchange, where it should have been doing business for the previous 100 years, if it hadn't been so stuffy. Of course, its timing was spot on, as it now had a big enough home to handle the expansion of trade brought on by the steam engine and the Industrial Revolution.

And thus began the modern insurance industry.

But to me, something is terribly wrong with the whole business structure. You would think that underwriters would hire gunships to protect the merchants they insure. But no, that task falls to the King and to Parliament, to collect taxes and custom duties on trade to pay for the Royal Navy to protect the merchant ships. The rewards, profits from trade or the upside of premiums, are privatized. But the risks are socialized. You might argue that the underwriter is bearing the risk of a ship being pirated or sinking, but he is getting paid for that risk. Everybody pays the cost of the Navy, through taxes, duties and forced conscription. Sure, every Englishman benefited from the strong Navy, but none more so than the underwriter.

In 1752, Ben Franklin helped charter the Philadelphia Contributorship for the Insurance of Houses from Loss by Fire. With people smoking that Virginia tobacco in bed and all, fires were a big problem in all cities. Fire departments were scarce, so insurance helped homeowners hedge their risk of getting smoked. But the lesson is that Ben Franklin and PCIHLF helped fund fire departments, to keep their payouts of Loss by Fire to a minimum. They were rewarded for privatizing risk. Special codings appeared on buildings to decide who was in charge of putting out the fire. There was a public uproar when some of these fire departments would watch a house burn down because they hadn't insured it.

So somewhere along the way Philadelphia, and every city big and small, established a municipal fire department, paid out of tax dollars. But Fire Insurance companies still charged premiums against Loss by Fire, perhaps not as much as before, but they still charged. Why didn't the Fire Department charge premiums, and get to keep them if they put the fire out before there was damage? Hmm, no good answer, is there?

• • •

James Watt and John Wilkinson and the Industrial Revolution did more than provide cheap, high-quality, nonitchy clothes to the world. They changed the entire support structure for families. Former farmers headed to cities to get higher-paying jobs at mills and factories. But that entailed risk. The worst thing that could happen to you on a farm, beyond the normal deadly 18th century diseases, was stepping into a pile of cow manure or having an ox step on your toe. And even if you were laid up for a while, the corn would still grow or your pregnant wife could probably milk the cows for a couple of days, no biggie.

But in these new-age factories, it was common to have your arm severed off in a Spinning Mule, or fall into a vat of molten iron ore, or become part of the pattern in a runaway loom. Since you and your family lived in the city, there was no backstop, no cows to produce milk or wheat in the field that neighbors might help harvest. If you died, your family was in grave danger of starving or freezing to death not soon thereafter.

These increasing numbers of deaths gave birth to the life insurance industry. Established in England and the United States to protect workingmen's families against disability or death, it was a noble offering. In the U.S., the Insurance Company of North America first began writing life insurance policies in 1794. Fraternal orders of the early 1800s, Freemasons or Knights of Columbus, for example, provided unofficial care for families of deceased brothers and over time these became premium-charging life insurance companies.

Again, life insurance companies provided a valuable public service, peace of mind for workers worried about their families. But beyond collecting premiums and paying out death benefits, they did little to make factories safe for the policy buyers. In fact, factory owners had little incentive to improve safety, since insurance seemed to cover their backs.

One huge problem was that many insurance companies were either scams or fiscally insolvent by the time it came to collect on death benefits. In 1851, New Hampshire became the first state to regulate insurance, and after the U.S. Supreme Court in 1869 ruled that insurance was local commerce and none of the federal government's business, the rest of the states followed.

Almost as soon as the automobile was invented, you guessed it, the auto liability insurance followed, and then policies insuring against auto damage. In 1905, a huge insurance scandal in New York spurred the passage of tough laws on how insurance is sold and money managed, much of the same regulation that exists today. This didn't stop insurance from rolling on.

In 1906, the San Francisco earthquake caused 500 deaths and $374 million in property damage, helping make the case for insurance against catastrophe. As industrialization continued to grow in the U.S., workers compensation laws were put in force in 1912, creating almost mandatory insurance against worker injuries. The U.S. involvement in World War I saw the creation of soldiers insurance, and even as the war was ending, a 1918 flu epidemic caused 100,000 deaths in three months.

The 1870s were a period of insurance mania; a railroad and telegraph boom meant a strong economy and workers got suckered into insurance for every possible tragic event. But as insurance proliferated, more money was sucked out of the higher-risk innovation economy, and instead put in safe, but low-return investments.

• • •

Insurance really got out of hand during the Great Depression in the 1930s, less to protect workers, and more to ensure that health care providers got paid for their services.

Health insurance had been around since 1850, when the Accidental Death Association of London, for an additional fee, offered to pay health benefits to workers who had severe injuries and couldn't work, in addition to death benefits. It has been a slippery slope since then. At the same time, the Franklin Health Assurance Company added similar coverage, again on a limited basis, for workers who could no longer generate income for their families. Of course, back then, typhoid and smallpox and all sorts of nasty diseases were circulating, helping to sell this health coverage.

For another 70-plus years, not that many people bought the idea, and most just paid for health care out of their savings. But then the Depression hit, and savings quickly disappeared to pay for food, let alone health care. This made doctors and hospitals quite nervous. How were they going to be paid for their services if no one had any savings?

In 1929, the Baylor Hospital in Dallas cut a deal with local schoolteachers. They would pay a monthly fee in exchange for potential services at the hospital. This became known as the Blue Cross plan. The teachers liked it since their health care costs were predetermined, and the hospital liked it since it guaranteed payment. The always envious doctors, not wanting to be left out of this whole "getting paid" thing, created a similar plan in the early 1930s named Blue Shield. People paid a monthly fee and the plan paid the doctors directly for any health care services they provided.

During World War II, when wages were frozen for workers, companies began offering health care insurance as a fringe benefit to attract workers. Tax laws helped; the government did not tax health care or other fringe benefits. Health insurance became a tax efficient way to pay workers.

The problem was that the whole industry—hospitals, doctors, insurance companies—got addicted to the insurance payment game. In fact, they used it to jack up prices and perhaps even sell unnecessary diagnostic and treatment services. Health care was 6% of gross domestic product in 1969 and 12% in 1987, and is easily pushing 20%. It's not managed care—it's managed payments.

Did health care get that much better? Maybe. Life expectancies have gone up, but maybe that's just vitamins and CPR. Who knows? Managed care, and HMOs and PPOs and all the other changes du jour to attempt to rein in medical expenses, don't seem to work. The money pooling and nonaccountability for runaway costs means that once again we are socializing the risks, and privatizing the rewards.

What else did the Depression bring us? Well, how about Social Security, insurance run by the government. The passing of the Social Security Act in 1935 mandated that workers contribute into a defined benefit retirement plan. Critics say it was a bribe from younger workers to pay off older workers so they'd retire and give up their hard to find jobs. Maybe so.

When the government in 1939 extended Social Security benefits to the surviving families of workers who died, it transformed the program into a life insurance policy. In 1950, if you worked for yourself instead of a company, you were covered too (of course, you had to pay up ahead of time, like everyone else). In 1956 it became a disability plan. In 1966 the government added Medicare and made it a health insurance plan. Of course, each of these came in addition to all the private insurance plans that workers already paid into for retirement, life insurance, disability, health care and who knows what else.

• • •

An outgrowth of Social Security was the Employee Retirement Income Security Act, ERISA, a bunch of gobbledy-gook government rules intended to protect retirees from having current and former employers steal their retirement money.

Taxpayers are the ultimate backstop for all forms of insurance. Take the Federal Deposit Insurance Corporation, which insures bank accounts up to $100,000. This allowed banks to take huge risks, and as a result, taxpayers had to bail out the savings and loan industry in the 1980s. Similarly, as pension plans at Enron and other companies go under, taxpayers, via the federal Pension Benefit Guaranty Corp, once again must bail them out. We end up with insurance on insurance on insurance.

There is a lot of talk of securitization in insurance, so companies can sell off their long-term liabilities to investors seeking that type of risk profile.

Even better, we could go back to the Lloyd's model, using your net worth as a backstop and collateral against risk. I would be willing to insure against hurricanes across a portfolio of homeowners in Florida if they would insure my home and others against earthquakes. The technology exists to match these types of liabilities, but the extreme state-by-state regulatory environment for insurance makes it unlikely for a long time. So it's back to the stock market to supply liquidity and risk capital. And that's exactly what it's good at. A $10 trillion valued stock market can probably provide deeper pockets of liquidity than the assets of Lloyd's of London "Names."

The Modern Stock Market

The modern computer helped create the modern stock market. In fact, they grew up together. The stock market provided the capital for the computer industry to create faster machines to handle Wall Street's own growing computing needs. Wall Street used to be a collection of people, but the permeation of computers into the business of raising and managing capital changed the way the business was done.

The modern-day Wall Street firm, which is barely even near Wall Street anymore, is not your typical command and control organization. There is no factory, barely any assets, just people making money helping others raise/manage/make money.

And what a funny collection of people it is. The B-school folks in expensive suits usually focus on underwriting; buying shares from a company at 4:30 P.M. Wednesday and placing it with investors by 9:00 A.M. Thursday morning. Not much risk, but investment banking fees can be as high as 7% for initial public offerings to 1–2% for follow-on stock sales, and on down. Success is based on a human network, relationships with companies, and a full court press when you smell a company about to raise capital. Once a company is public and it's stock is trading, bankers then work diligently to have someone buy shares, since this can mean $10 million to $50 million payments for merger and acquisition advice. Technology was never much help in banking, and the resulting barrage of phone calls and e-mails merely serves to annoy CFOs. Almost anything can be sold, although when there are too many firms chasing too few deals, fees tend to evaporate. So, to inflate fees, bankers learned to program and invent exotic derivative securities to pitch.

Bankers don't get their hands dirty selling or trading shares, there are cavernous trading floors to deal with that. Once a stock or bond is initially sold, it happily changes hands again and again. Commissions or spreads vary based on how easy it is to trade, or how exotic an instrument might be. The difference between making money and not is the flow, understanding who is selling and how much. A good trader works the phone to figure this out, a great one senses it from the trades flying by. It used to be said that to find the best traders with "street smarts," take a taxi into Queens until the meter reads $20 and then load up the back-seat. Today, the ticket to the dance requires an MBA. Typically, there is a division between bond trading and stock trading, or to make it sound important, fixed income and equity.

At first, Wall Street would just buy and sell shares as an agent for money managers like Chase Bank or Fidelity. But they noticed that these firms were making .8%–1.5% fees, year in and year out, for picking stocks. Not a bad business. So almost without exception, Wall Street firms have built asset management arms, and filled them with cerebral types, who either were good at picking stocks or got sent back to the trading floor. Of course, this has become yet another overbuilt business, and fees are constantly squeezed, so the only way to make money is pure scale; you must grow big enough to make it up in volume. These divisions are kept separate from the rest of the firm, to perpetuate the illusion that there is no conflict. They are not fooling anyone. In response, giant fund management company, Fidelity, tired of the effect of this conflict by those enacting their trades came up with a solution. Fidelity, which created its own brokerage firm to handle its trades and tossed Wall Street some bones to pay for research, remains the Street's largest customer.

Finally, and controversially, Wall Street makes money by proprietary trading, where firms simply trade for their own accounts. The easiest way to do that is to figure out what their customers are doing and then trade ahead of them. Subtly, of course. A more demanding way firms do this is by creating their own information advantage, finding profitable parts of the market to put their money into, and doing it in a big way. After NASA downsized in the early 1980s and froze hiring, Wall Street hired the hundreds of now unemployed pipe-smoking math PhDs—the rocket scientists— and locked them in back rooms to do quantitative research. They became known as Quants.

Morgan Stanley for years had a black box that would search out inefficient markets and arbitrage them to squeeze out some profit. When the black box stopped working, they fired the Quants and went back to stalking their customers' trades. After the biggest, baddest black box failed, the one belonging to now infamous hedge fund Long Term Capital, a postmortem found that Goldman Sachs and others had been mimicking LT Capital's trades with their own capital.

All this has started to change. Gigahertz PCs and gigabit data connections are changing the landscape. The "old way" of doing business is increasingly obsolete. Real machines are replacing people, not just in the back office, but also up front, in the profit centers. The structure of Wall Street will adapt to the new technology, although many, like the New York Stock Exchange, will go kicking and screaming. Let's take a look at how computers were introduced to Wall Street and how they slowly began to change Buttonwood-era ways of doing business.

From the Depression of the 1930s until the post–World War II industrial reconstruction, Wall Street went sideways. But never resting, the tech monster lurked. It wasn't until the late 1950s and 1960s that Wall Street emerged from its post–"Crash of 1929" slump.

In 1949, punch cards were used to record trades, and were then sorted by automated tabulators. Can you hear IBM salivating? In 1950, electronic computers started doing the sorting. In 1961, magnetic tape stored data. And in 1964, computers were used to clear certain trades by matching records. The post–World War II electronic computers from IBM and others that came out of those Moore School seminars were indispensable to Wall Street. Wall Street repaid the favor by bidding up IBM's stock to (indirectly) provide it with as much capital as it needed to grow. That's the stock market at its finest.

The go-go years on Wall Street from 1955 to 1973 provided a double bonus for Wall Street. First, stocks took off and trading volume increased. Since commissions were fixed, that increased volume drove profits straight up as well.

There were lots of limos driving Wall Street's titans around town. The hot growth stocks of the era were known as the Nifty Fifty. Names like Xerox, Polaroid and the rest were one-way growth stocks, and that way was up. The Vietnam War and LBJ's Great Society needed to be financed.

An era of deficit spending saw Wall Street create a liquid bond market to fund the government's increased debt. The government would sell notes, bills and bonds to Wall Street, which would, you guessed it, mark-'em-up and unload them on customers. Bond spreads were as wide as a country mile, and banks, insurance companies and pension funds were loading up and paying Wall Street to do the trades. In fact, they had no other choice.

As volume and profits grew, more and more Wall Street partnerships were created to get in on the game. Brokers and traders grew in number, and they all feasted on fat commissions. It was a people business. When institutions or individuals wanted to buy a stock, they called their brokers who relayed the order down to the floor of the exchange. Electronic message switchboards replaced the pneumatic tubes of 1918 so order entry and execution confirmation could happen in almost real time. Of course, this was to the advantage of the traders; they could get the most current feel for the market and customers were still out of that loop. But the back office was neglected and stock certificates still had to be collected and distributed. Wall Street hit a wall in 1968 when electronics facilitated trading but did nothing to help clearing. For many firms and partnerships that was fatal.

As volume increased by some 30% a year in 1967 and 1968, the people-intensive clearing process took forever. Certificates piled up to the ceiling tiles at brokerage firms, either awaiting payment or as errors, bad trades, or yet to be reconciled. Hiring more people barely helped. Firms would work 24 hours a day, 7 days a week and still fall behind. When customers realized they weren't getting their certificates they simply stopped paying. The exchange began closing one day a week to catch up. The NYSE formed a Central Certificate Service that created electronic certificates for a few stocks so settlement and clearing could happen without handling physical certificates. It was too little, too late. In 1969 and 1970, it took so long to process the volume of trades that 160 NYSE member firms went belly-up, their credit squeezed.

Small companies, who couldn't afford the NYSE or didn't have the size or pristine balance sheets needed to be listed, could have their stocks traded off-exchange, or OTC, for over-the-counter. Until 1961, it was a pretty sleazy business, trading was thin, quotes were by appointment, and spreads were at the whim of one or two traders. In 1961, the SEC empowered the National Association of Security Dealers, or NASD, to automate the trading of these OTC stocks. Pretty impressive for 1961. Unfortunately, it took until 1971 for the first NAS-DAQ stock to trade, as lots of new technology needed to be invented. But it created another huge market for all these new-fangled computers.

Beyond the NYSE and NASDAQ, computers were prolif-erating everywhere. Order handling and accounting were big paper problems for America's corporations as well. New and funky technology companies arose to solve problems similar to Wall Street's paper crunch. Wall Street was all too happy to funnel risk capital to these companies—IBM and the BUNCH: Burroughs, Sperry-Univac, NCR, Control Data, and Honeywell.

But raising capital on Wall Street was typically limited to blue-chip companies and merely required a handful of phone calls. "We are raising $200 million for a new line of yellow cheese for Kraft, how many shares can I write you down for?" Tech deals were almost impossible to sell that way. Understanding Chips Ahoy! is easy. Silicon chips—that's harder to explain.

In 1971, C. E. Unterberg Towbin had one particular client with a very hard to explain technology, strange devices named I2102s, which were 1 kilobit of static memory made with a planar process, and that were rapidly replacing core memory in IBM computers. This company also had a new "micro" processor used in some Japanese calculators. Yup, Intel. The phone didn't cut it, so the company's execs trooped out to the offices of institutional investors across the country to sell their IPO. The road show was thus invented. Now all companies have to go through this grueling ritual. Not so much that investors demand the equal access, it's simply that the only way they can be sold on some new, complicated idea is to have it explained to them face-to-face and with diagrams. Wall Street was all too happy to take these new companies on the road—IPOs paid 7% fees, and still do.

The newly financed technologies went to good use. In May 1973, IBM, for example, by using cheaper Intel memory, was able to create a system of electronic record transfer of ownership for the Depository Trust Company, or DTC, a new entity that had taken over the clearing process. That allowed 16 million shares to trade every day and shares from 4,729 different companies were registered with the DTC.

With DTC, the back office paper crunch ended. No more lost business from closure. With an automated back office, the risk of investing on Wall Street went down. This is a subtle and critical point, but lower risk on Wall Street meant more risk capital could be deployed elsewhere. This meant bigger risks could be made away from Wall Street, investing in the Intels of the world.

The economic boom meant that the amount of money under management ballooned and Wall Street's volume and profits exploded. But the institutional investors' annoyance with fixed-rate commissions grew as well. In 1971, New York State Controller Arthur Levitt, whose son Arthur Jr. would later become chairman of the Securities and Exchange Commission, was trustee of the state's Common Retirement Fund and was tired of paying huge commissions to move in and out of stocks. So he demanded membership in the NYSE so he could, in effect, get a rebate on the painful commission rates.

The NYSE of course refused, so he took his problem to Washington. Reform was in the air after a nasty bear market of 1974, when the Nifty Fifty turned shifty, and lost a fair amount of value. The Securities Exchange Act of 1975 did away with fixed commissions on trading, and mandated an electronic National Market System. The so-called Big Bang would expose the fat cats on Wall Street to the full winds of technology-induced change and that was to be the end of them. Some, like a firm called White Weld, disappeared, mainly by merging into bigger firms. The rest adapted.

With most of the weak players wiped out from the paper crunch in 1969, and systems like DTC in place to handle the volume, Wall Street firms adapted to lower commissions by putting up their own capital to facilitate trades and getting paid for their liquidity. Putting up money substituted for thinning spreads from the end of fixed commissions. Wall Street legend tells of a sign on Goldman Sachs's football field–sized trading floor that read: "If you can't find the other side of the trade, you are it."

When I hung out at trading desks, I would hear this all the time: "Hi, I've got a million shares of Philip Morris for sale at $54. Can you move it? I'll offer you $53¾ for 200,000 shares or $53½ for the million." Then the trader would scramble around to find lots of buyers at $54 to $54¼, focusing on the usual suspects—Fidelity, Morgan, Putnam—these guys always needed something. Then he would "cross" the million-share block, and make $500,000 in a few short minutes, a pretty good trick at 6 cents a share.

The National Market System moved ahead and created the NASDAQ trading system for electronically trading OTC stocks, which were all the newly public, high-growth, but "risky" companies. Its first big test was in 1983. Personal computers and video games were sucking up lots of chips, and dozens of companies in Silicon Valley were anxious to raise capital. The IPO market rolled over in 1984, but came back in 1986 when Microsoft and Compaq went public. Goldman Sachs and Morgan Stanley made big underwriting and trading fees when they took these firms public. Ten years later, PCs were big contributors to the eventual demise of Wall Street's trading profits. But hold that thought.

It's pretty obvious that it was this new electronic NAS-DAQ, and not the stodgy New York Stock Exchange that became the forum for raising risk capital, to fund the next wave of great American technology companies. Nonetheless, the rules and regulations are still designed to protect the stodgy boys' cozy trading world, and provide the biggest impediment to funding new ventures in the U.S.

Wall Street has always salivated over small investors, the suckers who pay big spreads without realizing that they are getting ripped off. In 1985, the NASDAQ programming wizards cooked up a system called Small Order Execution System, or SOES, so small investors could automatically get trades of 1,000 shares or less executed. Almost no Wall Street firm used it. Why bother? Why automate a process when someone on the phone could squeeze out a bigger spread?

But on October 17, 1987, known as Black Monday, the markets crashed, and traders refused to answer the phone. I know. I was working at PaineWebber as a research analyst at the time. With nothing better to do than watch my career disintegrate with each dip of the Dow, I wandered out to the trading floor and watched the chaos for a few hours. The phones rang off the hook. A friend who was a trader explained that if he answered it, it was most likely a sell order he couldn't fill, at any price. So screw it, let it ring. This happened up and down the Street that afternoon.

Bad move. The system itself held up just fine, 604 million shares traded, almost double the old record of 338 million shares, set in the previous trading session the Friday before. But with the Dow down 22.6% in a day, congressional hearings were inevitable. "Aren't there rules to prevent this? No? Then we need more rules."

One of the complaints was the inability of individual investors to connect with brokers or traders. Someone let slip out that there was indeed this cool system named SOES that automatically executed trades of 1,000 shares or less. "Well, how did it work on Black Monday?" "Uh, it wasn't really turned on."

In late 1987, all Wall Street firms were ordered to turn on the SOES and USE IT. As usual, the law of unintended consequences took hold. The SOES rule ended up helping very few small investors because traders just dropped trading many stocks to avoid the risk of getting picked off for 1,000 shares. But over time, for a very different reason, SOES ended up obliterating profits for NASDAQ traders.

Since spreads on most NASDAQ stocks were 25 to 50

cents, there was lots of room for someone to come in and trade "inside" the quoted spreads. A young guy named Jeff Citron at a small firm named Datek and his programmer friend Josh Levine quickly figured out they could use the SOES system and eat the big guys' lunch. I met Levine briefly, and he launched into the benefits of blood oranges. Programmers are like that.

Writing code using MS-DOS, Levine created a program that would find bid and ask spreads on stocks that were wide enough for his liking and then with the click of a mouse, fire off a series of 1,000-share orders using SOES and pick off Wall Street traders. He generated profits with almost no risk. No longer did you need huge stakes of capital to make money trading. It cost $2,000 for a PC, and probably $10,000 up front to get the first trades settled.

A five-minute rule on SOES prevented a trader from doing the same trade with the same stock within a five-minute period. In addition, anyone doing more than five trades a day was considered a professional trader, not a retail customer, and could be thrown off of SOES. So to stay within those rules Citron created a brokerage company, and invited as many individuals as he could to plug in and use his SOES system, creating an army of fast moving traders. Individuals had to wait five minutes, but the collection of them was like a Tommy gun. Citron invented Day Traders.

These guys became known as SOES bandits, and their effect was to decrease liquidity for NASDAQ stocks, as firms would no longer put up capital for their clients. Why facilitate a trade for JPMorgan or for John Q. Public if the SOES leeches are going suck you dry before the trade was completed? A few years back, I stopped in New York to see one of my buddies, a NASDAQ trader with whom I did a lot of business. He stepped away from his terminal for a few minutes to chat, always a dangerous move, but his system kept chiming, and a series of red flashes scrolled along the bottom of his screen. He noticed me looking, and just rolled his eyes and said "SOES bandits," and then threw me out and went back to his screen.

Traders still had to trade. When they got a big order, it was common for a trader at one firm to communicate their intentions to cross a large block of shares with traders elsewhere. The rules state that traders must trade at the prevailing best prices. Just one bid at an eighth higher meant he couldn't trade the block, so he basically told everyone else, in not so many words, to get the heck out of his way, as he needed the best bid or ask at the time to trade the shares. Smart traders work around dumb rules. This very action is what led to a 1996 class action lawsuit by securities litigation lawyer William Lerach alleging collusion and spread fixing among NASDAQ market makers, which eventually was settled for a cool $1 billion. This also led the SEC to enact Rule 11Ac1-4, the Limit Order Display Rule. And the dominoes started to fall.

This 1996 rule effectively enabled Electronic Communications Networks, a fancy name for a system that electronically matches orders. A buyer and a seller agree on a price electronically rather than via the phone. No brokers, primary buyer to primary seller. It was anonymous and it worked in milliseconds, a blink of an eye. Instinet was the original online matching service, but it was of little value until it could see the Street's limit orders, where customers were willing to trade above and below the current spread. Though it started in 1968, and was bought in 1987 by Reuters, which had smartly moved beyond carrier pigeons, this electronic matching service saw volume take off in 1997.

Josh Levine, the programmer for SOES bandits, went back and tweaked his MS-DOS code and quickly created his very own ECN, naming it The Island. With Wall Street withdrawing liquidity from NASDAQ trades, and unwilling and perhaps unable to cross shares at a spread it could still make money with, The Island ECN came along as simply a matching service. A Web interface on the front end and an industrial strength matching system on the back end, his system quickly ate into not only Instinet's bread-and-butter business, but also all of Wall Street's NASDAQ volume. By the end of 1998, ECNs accounted for 21% of NASDAQ volume. Then 26% in 1999, 36% in 2000 and close to 50% of NASDAQ volume was done by ECNs by the end of 2001. It is now over half and growing.

Those traders should have answered their phones. A big chunk of Wall Street's small stock trading profits disappeared.

Until 1975, technology was needed to save the back office and its clearing problems, but Wall Street enjoyed a period of fixed commissions and increasing volume. Big Bang meant commissions and spreads dropped, but volume increased. Large Wall Street firms provided liquidity and their own capital to differentiate themselves. They quickly made up for declining commissions.

The 1980s saw trading volumes increase, and a bull market sure helps. Big Wall Street firms made money hand over fist. But by the mid-1990s, spreads were shot to hell by SOES bandits and day traders. Liquidity couldn't save the day. In fact, it was a surefire way to lose money. Goldman Sachs took down its famous sign, and probably replaced it with one that reads: "If you are the other side of trade, you're fired."

A Web interface meant you could trade stocks without

trading phone calls with traders, and from a hotel room or a hot tub.

The biggest concern about using an ECN to trade was liquidity, i.e. whether there were enough participants plugged into the same system to find others to trade with. Unlike Wall Street putting up capital to provide liquidity, size equated with liquidity. The community of users provided a liquid market, maybe 100 shares at a time, but liquidity nonetheless. Think eBay or Instant Messaging, worthless with a handful of users, but unspeakably powerful platforms with millions of users. Metcalfe's Law at work! The number of users matters, and trumps whatever capital a Wall Street firm could throw at the problem.

While NASDAQ volumes rose with cheaper trading via ECNs, the Big Board felt threatened. Traders there clung dearly to that "no trading outside the exchange" culture. Barely 10–15% of Big Board volume was done off exchange. A centralized exchange when technology is decentralizing everything in its path doesn't stand a chance. But never underestimate the power of a regulated and de facto monopoly business. As an exchange, the Securities and Exchange Commission blesses its rules and regulations. It ended its fixed commissions in 1975, but didn't break up its monopoly.

Still, spreads were under pressure and it every year became tougher for specialists to make money. In a massive changeover, 1999 saw the Big Board convert to dollars and cents instead of dollars and fractions. This decimalization, quoting stocks down to the penny instead of quarters and eighths made things worse. On the surface, this seemed like real progress and a problem only for specialists, who instead of being able to trade with one-eighth or 12.5-cent spreads, had to move quickly to 5-cent spreads, and in really high-volume stocks, those that trade for as little as a penny or two. But no specialists complained, not seriously anyway. Why? Quite simple, specialists now had a license to steal, to trade ahead of their customers for an added penny instead of costing them 12.5 cents.

However they pulled it off, being a specialist was a profitable business, and this might explain why specialists have consolidated and have been snapped up by Wall Street firms. There were 57 specialist firms in 1986; at the end of 2001 there were nine.

In 2004, NYSE chairman Dick Grasso resigned in a flap over his pay. He made some $200 million over a period of years, which many found excessive. I suppose the real issue was an exchange chairman both setting the rules and making money from the rules. He also helped the NYSE fight off automation via a sneaky little, misunderstood rule known as the trade-through rule.

Like it or not, how money is managed has changed. Gone are the days of big ugly banks and massive pension funds buying only blue chips. A new style of investing has emerged that is fast and furious, buying and selling baskets of hot chips, if you will. Unlike buying industrial dinosaurs, this style encourages both the formation of risk capital for entrepreneurs, and an exit strategy for venture capitalists that fund entrepreneurs.

No three-martini lunches for this crowd. Managers at these funds work at a frenetic pace, always looking for the latest and greatest. They are willing to invest in higher-risk businesses, if they think there is fast trading execution and enough liquidity for them to get out if they perceive things might soon get dicey.

I know this sounds bizarre, but this style has a discipline

that ends up allocating more capital to risky ventures with big potential than Wall Street ever has in the past.

With ECNs and this new rapid-fire system of trading, a subtle shift began on how Wall Street managed money. Fast trades meant accounts could turn over their funds more. On the surface, this seems like a threat to capital formation, in other words, the death of long-term investing. But perversely, this new fast and furious approach meant growth fund managers and hedge fund managers could try new things, and invest in new markets without the fear of getting stuck with shares if the investment went bad. MORE capital was available for riskier, high-growth companies, not less as would seem intuitive. Automated trading meant more risk capital.

These new fund managers take risks, with assurance that the companies they invest in provide accurate information, have liquid shares, trade cheaply and quickly and exist free of stock manipulation. These funds rarely own Russian gas refiners, as they fail all of the above assurances. But funds do own weird companies that make components for optical wave division multiplexing or some new biopharma company or the latest in supply chain management software, but they require automated trading systems to stay quick of foot.

The New York Stock Exchange is still a people-intensive exchange. Its specialist system was created in 1871. And we are still stuck with the NYSE monopoly on listed shares. Lots of reasons are offered, such as centralized pools of liquidity or orderly markets, etc.

But also to blame are some subtle regulatory technicalities that the NYSE hides behind. One technicality is a network called the Intermarket Trading System, which was also set up in 1975, ostensibly to execute trades between exchanges. And, you guessed it, only registered exchanges can plug into it directly. But it's worse than just not having access. An ECN can't just match a listed trade. A "trade-through rule" states that if you want to trade a listed stock at a price "inferior" to the price advertised by the specialist at the exchange, you have the obligation to go to the exchange first. An order is not placed through this system, it is requested as a "commitment to trade." This trade-through rule gives an exchange 30 seconds (it used to be 60) to respond, and it's a one-way commitment, the specialist can just ignore it.

It's lose-lose for the ECN. They can't wait 30 seconds, markets move too fast. But if an ECN doesn't bother to contact the specialist and instead trades through, the specialist can insist on a piece of the trade, a liability ECNs are not set up for. So very little listed share trading is done.

You may ask, why would a customer do a trade at an "inferior" price, and not just go down to the exchange? Because quotes can be stale or not represent much volume, the term "inferior" can be deceptive. The rule was set up for humans, not computers.

In trading, 30 seconds is an eternity, and there is no way an ECN (which typically executes trades in .03 seconds) can guarantee best execution when 30-second delays are built into the system.

On November 28, 2001, when Enron's stock finally gave up the ghost, it opened at $3.69 and closed at $1.10. There were so many sell orders that the specialist at the NYSE, whose sole responsibility is to maintain an orderly market, declared an order imbalance, effectively halting trading with the stock stuck at $2.80 per share. The trade-through rule goes away if the exchange is closed, so according to Archipelago, 10 million shares of Enron were matched on ECNs, at prices between $2.80 and $1.20. The specialist at the NYSE finally reopened the stock at $1.20. Who needs him?

Good question. To find out I took a tour of the floor of the NYSE at Wall and Broad. I found myself in front of the station where shares of Micron Technology, a company I have visited in Boise, Idaho, trades. There were a couple of floor brokers hovering around a bored-looking specialist. Some were trying to sell, others just checking out the action or the female anchor on CNBC, while the specialist occasionally knocked down the price to see if the seller would go away. Except for the TV coverage, you could find the entire setup duplicated in a Web-based software applet on ECNs.

I met with Bob Britz, an executive VP who was exceptionally bright but a little overenamored with the NYSE setup. I would have liked to have met with Chairman Dick Grasso but he was busy. I looked up at some commotion in the balcony at the market close to see Dick Grasso and Mr. Potato Head applauding the closing bell. It was Toy Fair week in New York but I must not have made the cut.

I suggested to Bob Britz that if the trade-through rule disappeared, ECNs would take 50% of his volume, as they did with NASDAQ. Of course, he couldn't agree. His answer was circular: "So as long as we have the market share and liquidity, we will keep it [the market share and liquidity]".

Except as far as I can tell, each of the users of ECNs provide their own little contribution of liquidity. Each one may be small, but en masse, as a community of liquidity, they blow away the liquidity of any single specialist, and probably in total more than any single investor. My sense is the 200-year-old NYSE would rather not test my theory, but instead continue to hide behind its 25-year-old regulations.

What's at stake? Specialists make huge profits from trading, for clients and for themselves.

So we are back to thinking about risk capital. Excessive fees in capital markets raises the cost of capital, and as a hidden tax, lowers long-term growth rates. Penn State University Professor Ian Domowitz studied global markets and noted that buying listed shares costs 429% more via traditional brokers than electronic venues. And he also calculated that a fund with twice a year turnover and a 12% annual return would have listed exchange-trading costs eating up 23% of returns. That is a heck of a tax just to keep the quaint NYSE going.

Remove a few technical regulations to enable automated trading, and within three years, half of NYSE listed share trades would be done by ECNs, and the Buttonwood Agreement would finally be laid to rest.

It took a couple of hundred years, but a new structure is emerging for Wall Street that can fund an idea economy instead of an old industrial economy that might quickly become irrelevant.

The U.S. can continue to grow as great new companies can be formed on ideas instead of assets.

The U.S. has the ultimate tool—a vibrant stock market that sets prices for companies as well as helps spread risk to an acceptable level. I've heard it referred to as risk pooling. Entrepreneurs take the ultimate risk. But via liquid markets, investors and their capital are more willing to take on risk. Liquidity doesn't limit the downside, it merely provides the comfort that if an investor's mind changes on a company or a sector, they can get out, and fast. In other words, the ultimate stock market is like a dial tone—it's just there and executes trades. But this only goes so far. If investors believe that the stock market is a hindrance, that trading through a specialist at the floor of an exchange adds friction, they are less willing to make riskier investments. In other words, the stock market can't be part of the risk, or capital goes elsewhere. This is the problem you find in foreign exchanges, and why it is so important that the U.S. keep its capital markets the fastest, the cheapest and the most efficient. If it doesn't, capital will slosh away.

Of course, what comes with all of this is volatility. So what? A fool and his money are soon parted. Speculation and panics will be part of the system. Remember, the emergence of the American industrial machine saw investors chase up stocks and then dump them just as quickly when the bubble burst. You can't stop people who get overextended with leverage and fail—it's inevitable, get used to it. As the industrial age dies and the digital age takes over, expect many more blowups. The Internet Bubble of 1999 was just a first step. But if, unlike anywhere else in the world, the U.S. makes sure its markets are structured to enable the formation of risk capital, the arrow, while jagged, will be up and to the right over the long run. Entrepreneurs and their ideas will have the proper funding to become a reality.

So just as steam-driven textile mills provided cheap and comfortable clothes, and microprocessors provided cheap and powerful computers, new technology is displacing old ways of doing business. The Buttonwood-era stock market is now an obsolete institution, and being replaced by matching servers sitting in dark rooms in Jersey City. Wall Street will never be the same. But the good news is these millisecond-matching machines will facilitate more risk capital to fund innovation.

Just the existence of speed stimulates liquidity, and that is the ultimate tool of the stock market to efficiently allocate capital. Pascal didn't get funded by stock markets, and neither did Boulton and Watt. But Edison got capital, as did Intel and Cisco and now Google. The pace of innovation is quicker with capital to fund it. To paraphrase Woody Allen, Technology and Wall Street can lie down together, but neither will get much sleep.

ENIAC Press Release

WAR DEPARTMENT
Bureau of Public Relations
PRESS BRANCH
Tel. - RE 6700
Brs. 3425 and 4860

F U T U R E

FOR RELEASE SATURDAY A.M. , FEBRUARY 16, 1946

For Radio Broadcast after
7:00 P.M. , EST, February 15, 1946

A new machine that is expected to revolutionize the mathematics of engineering and change many of our industrial design methods was announced today by the War Department.

Designed and constructed for the Ordnance Department at the Moore School of Electrical Engineering of the University of Pennsylvania by a pioneering group of Moore School experts, this machine is the first all-electronic general purpose computer ever developed. It is capable of solving many technical and scientific problems so complex and difficult that all previousmethods of solution were considered impractical.

This mathematical robot, known as the ENIAC (Electronic Numerical Integrator and Computer), is the invention of Dr. J. W. Mauchly and Mr. J. Presper Eckert, Jr., both of the Moore School. Begun in 1943 at the request of the Ordnance Department to break a mathematical bottleneck in ballistic research, its peacetime uses extend to all branches of scientific and engineering work.

The ENIAC is capable of computing 1000 times faster than the most advanced general-purpose calculating machine previously built. The electronic methods of computing used in the ENIAC make it possible to solve in hours problems which would take years on a mechanical machine--a time so long as to make such work impractical.

Containing close to 18,000 vacuum tubes in its mechanism, the new machine is a giant of electronic precision. It occupies a room 30 by 50 feet and weighs 30 tons.

Although the machine was originally developed to compute lengthy and complicated firing and bombing tables for vital ordnance equipment, it will solve equally complex peacetime problems such as nuclear physics, aerodynamics and scientific weather prediction.

Official Army sources made it known that research laboratories of several large industrial firms have expressed active interest in the machine. These include manufacturers of electron tubes, jet engines, gas turbines, and other types of engines. Spending vast sums of money yearly for experimentation and design research on their products, these firms are naturally interested in any means of reducing such costs. It is further felt that better, more scientific design will now be possible, as a result of the new machine's facility for handling hundreds of different factors in one computation. Much lengthy and costly design experimentation, often involving the construction of a seies of expensive models, a common practice in airplane design, might also be eliminated. It was explained that such trial-and-error methods were successful in Edison's and Marconi's day, but would not suffice to deal with complex phenomena arising, for instance, from the blast of an atomic bomb.

The new machine does not remove the need for legitimate experimentation, whose purpose it is to discover fundamental principles and factors which affect these principles. Likewise, it was pointed out that the electronic calculator does not

MORE

replace original human thinking, but rather frees scientific thought from the drud-
gery of lengthy calculating work. The work of the English physicist, Dr. D. R.
Hartree, who spent 15 years in calculations on the structure of the atom, was given
as an example of lengthy computation attendant on modern science.

Cost estimates of the ENIAC run to about $400,000. This includes all research
and development work; future machines of this type could be produced much more
cheaply.

Built at the Moore School of Electrical Engineering of the University of Pennsylvania for the Amy Ordnance Ballistic Research Laboratory at Aberdeen, the ENIAC
is the result of a fortunate combination of men and circumstances. The origininal
idea for the electronic general purpose calculator was that of Dr. John W. Mauchly,
of the Moore School faculty. Dr. Mauchly, previously faced with many physical and
meteorological problems requiring voluminous calculation, had conceived electronic
devices for handling large computing problems in mass-production style. Captain
Herman Goldstine, mathematician and ballistic expert for Army Ordnance, saw in
these plans a powerful tool needed by the Ballistic Research Laboratory, which was
confronted with overwhelming computational work, and he enthusiastically promoted
the interest of the Ordnance Department in undertaking development of the ENIAC.
Administrative supervision of the project was assumed by Dr. J. G. Brainerd, of
the Moore School. Mr. J. Presper Eckert, Jr., a recent graduate of the Moore
School, joined with Dr. Mauchly in elaborating the plans for the ENIAC and took
charge of the technical and engineering work. Mr. Eckert and Dr. Mauchly have to
their credit a number of inventions relating to improved methods of electronic
computing.

The ENIAC was begun in July, 1943, and finished in the Fall of 1945. That a
development of this magnitude was completed in so short a time is largely due to
the close cooperation between Amy Ordnance and the Moore School. The appointment
of Captain Goldstine to maintain technical liaison was an important factor.

The ENIAC was sponsored by Colonel Paul. N. Gillon, a 38-year-old West Pointer
with a graduate degree from Massachusetts Institute of Technology. He has given
his enthusiastic support through the trying period which characterizes all pioneering developments in science. Colonel Gillon, when at the Ballistic Research Laboratory, was first interested in the project by Captain Goldstine. A transfer to
Office, Chief of Ordnance, enabled Colonel Gillon to give the necessary whole-
hearted support to the project. He sums up his role in this connection by saying
his part was "confined to such things as decisions on scope and flexibility; overall supervision and support; and fighting off the competition of apparently more
urgent but actually less important conflicting projects through the war period."

The designers of the ENIAC speak of it as a "digital" or "discrete variable"
computing machine, as opposed to the "continuous variable" type of machine, of
which the differential analyzer is an outstanding example. The latter are devices
that can handle only a restricted class of problems. But it was the experience of
the sponsors of the ENIAC that modern physical problems could no longer be handled
adequately by existing types of calculators.

Early in the emergency, the non-existence of ultra-rapid calculating facilities was recognized as a potential bottleneck in the swift delivery of vital ordnance materiel to troops in event of war. Artillery, for instance, after being
built and tested, cannot be put into use until firing tables, indispensable to
the operation of the guns, are supplied. Consequently, combat equipment could
pile up at shipping depots, waiting for the proper firing tables. That this situation did not occur was due to the enormous effort that the Ordnance Department
expended in utilizing all possible computational arrangements in facilities and
personnel to prevent it. The Ordnance Department had its own differential analyzer, in addition to the one it was using at the University of Pennsylvania. These
analyzers performed many labor-saving calculations up to a certain point
beyond that, a staff of about 200 trained conmputers was needed to complete the
tables. A general purpose digital type of machine was needed, so that all calculating operations could be performed by machine. The construction of such a machine was being undertaken by another institution, but, because of its mechanical
nature, its speed was nowhere near as high as that desired. At this point the
team of Eckert-Goldstine-Mauchly, under Colonel Gillon's supervision, began to
function and work commenced on the all-electronic general purpose computing
machine, the ENIAC.   - 2 -   MORE

The speed of this computer is phenomenal. The first problem put on the ENIAC, which would have required 100 man-years of trained computer's work, was completed in two weeks--of which two hours was actual electronic computing time, and the remaining time devoted to review of the results and details of operation. If used to complete capacity, the ENIAC will carry out in five minutes more than ten million additions or subtractions of ten-figure numbers. The machine performs a simple addition in 1/5000 of a second (and can do a number of distinct additions simultaneously); a single multiplication by a ten-digit multiplier in 1/360 of a second; a nine-digit result in division or square rooting in 1/38 of a second.

It is felt that the tremendous speed and flexibility of the machine will permit many immediate industrial uses, as well as application to far-reaching investigations of natural phenomena. Although the cost of such a machine might seem fairly high, value received in elimination of expensive design processes would easily cover the initial investment in a short period of time. Many electrical manufacturing firms, for instance, spend many thousands of dollars yearly in building "analogy" circuits when designing equipment. A continuous-type of calculator used in this work is the "network analyzer." Like the differential analyzer, it can solve only certain restricted types of problems and then only up to a certain point.

Sponsors of the ENIAC point out that it can carry out numerous "logical" operations but that it cannot do creative thinking. The mathematician, physicist or engineer is still needed--in fact, more than ever, to analyze the problem mathematically and set up the sequence of operations in the machine. It is expected, moreover, that machines of this type will bring a greater "mathematics consciousness" to engineering and production.

In the latter connection several large organizations, whose work involves much statistical work, have expressed interest in the ENIAC. Among these are the National Bureau of Standards and the Army Air Forces Weather Service.

A very successful place for the new machine is expected in the science of weather forecasting. Amy and Navy officials are most interested in this application. Instead of makeshift techniques, the machine would make possible the scientific analysis of the large mass of meteorological data that has been collected over the years. Not only would long-range accurate weather predictions be made possible, but weather in any spot in a given area could be determined almost instantaneously, when the "boundary data" are known.

- 3 -

DISTRIBUTION: Aa, Af, B, Da, Dd. Dm N, Sc.
1-30-46

Baldwin, Neil. *Edison: Inventing the Century.* New York: Hyperion, 1995.

Bernstein, Peter L. *Against the Gods: The Remarkable Story of Risk.* New York: Wiley, 1996.

Buck, James E. *The New York Stock Exchange: Another Century.* Greenwich Publishing Group, 1999.

Burke, James. *The Day the Universe Changed: How Galileo's Telescope Changed the Truth—and Other Events in History That Dramatically Altered Our Understanding of the World.* New York: Back Bay Books, 1995.

Ceruzzi, Paul E. *A History of Modern Computing.* Boston: MIT Press, 1998.

Ferguson, Niall. *The Cash Nexus: Money and Power in the Modern World 1700–2000.* New York: Basic Books, 2001.

Gilder, George. *Telecosm: How Infinite Bandwidth Will Revolutionize Our World.* New York: Free Press, 2000.

Google, 2001–2004.

Gordon, John Steele. *The Business of America: Tales from the Marketplace—American Enterprise from the Settling of New England to the Breakup of AT&T.* New York: Walker Publishing, 2001.

Grove, Andrew S. *Only the Paranoid Survive: How to Exploit the Crisis Points That Challenge Every Company and Career.* New York: Currency Doubleday, 1996.

James, Lawrence. *The Rise and Fall of the British Empire.* New York: St. Martin's Press, 1994.

Johnnes, Jill. *Empires of Light: Edison, Tesla, Westinghouse, and the Race to Electrify the World.* New York: Random House, 2003.

Landes, David S. *The Wealth and Poverty of Nations: Why Some Are So Rich and Some So Poor.* New York: W. W. Norton, 1999.

———. *The Unbound Prometheus: Technological Change and Industrial Development in Western Europe from 1750 to the Present.* New York: Cambridge University Press, 1969.

Quinlan, Joseph P. *Global Engagement: How American Companies Really Compete in the Global Economy.* New York: Contemporary Books, 2001.

Standage, Tom. *The Victorian Internet: The Remarkable Story of the Telegraph and the Nineteenth Century On-Line Pioneers.* New York: Berkeley, 1998.

Sylla, Richard. "The First Great IPO." *Financial History Magazine* 64.

Thurston, Robert. *A History of the Growth of the Steam-Engine.* New York: D. Appleton and Company, 1878.

Weightman, Gavin. *Signor Marconi's Magic Box: The Most Remarkable Invention of the 19th Century and the Amateur Inventor Whose Genius Sparked a Revolution.* New York: Da Capo Press, 2003.

Index

Andy Kessler is a former electrical engineer turned Wall Street analyst turned investment banker turned venture capitalist turned hedge fund manager turned writer and author.

His book *Wall Street Meat: My Narrow Escape from the Stock Market Grinder* was published in March of 2003. The follow-on, *Running Money: Hedge Fund Honchos, Monster Markets and My Hunt for the Big Score,* was published by HarperCollins in September of 2004.

Andy is a frequent contributor to the *Wall Street Journal* op-ed page and has written for *Forbes* magazine, *Wired,* the *Los Angeles Times,* and the *American Spectator* magazine as well as techcentralstation.com and thestreet.com Web sites. He has even written a piece of fiction for *Slate*—bet you can't find it.

Andy Kessler was cofounder and president of Velocity Capital Management, an investment firm based in Palo Alto, California, that provided funding for private and public technology and communications companies. Private investments included Real Networks, Inktomi, Alteon WebSystems, Centillium and Silicon Image.

In the early 1980s, Andy spent five years at AT&T Bell Labs as a chip designer, programmer, and spender of millions in regulated last-minute, use-it-or-lose-it budget funds. In 1985, he joined PaineWebber in New York, where he did research on the electronics and semiconductor industry and was an "All Star" analyst in the Institutional Investor poll.

In 1989, Andy joined Morgan Stanley as their semiconductor analyst, and following in the footsteps of Ben Rosen, he added the role of technology strategist and helped identify long-term, secular trends in technology. In 1993, he moved to San Francisco to join Unterberg Harris, where he ran a private interactive media venture fund, with investments that included N2K, Exodus and Tut Systems.

Andy received a BS in Electrical Engineering from Cornell University in 1980 and an MSEE from the University of Illinois in 1981. K-12 was at Bridgewater-Raritan High School East in New Jersey. Every morning for 13 years, while heading out for the school bus, Andy looked to his left, up the hill, and checked out the flag flying at Middlebrook Encampment, where George Washington and his troops spent winters watching the British troops in New Brunswick. On June 14, 1777, the Continental Congress approved the Betsy Ross 13-star flag as the official flag, and it flew for the first time at the Middlebrook Encampment. Pretty cool.

He lives with his wife and four sons in the Bay Area and enjoys basketball, hiking, skiing, biking, Pininfarina-designed moving objects and reminiscing about raising Siberian Huskies.

www.andykessler.com